This book is a useful resource for the implementation of the Mandatory Building Inspection Scheme (MBIS) which took effect in June 2012. The information on the ageing private building stock and the projected demand and supply of professional workforce are conducive to the effective implementation of the Scheme. At the same time, the research findings provide important insights into the urban decay and building safety problems in Hong Kong.

I wish to congratulate Prof. Leung and his team on their good effort in helping forge ahead the MBIS for the well-being of the community, and would recommend this book to all those concerned or interested in the issue.

Sr. WU Moon Hoi, Marco, JP
Chairman, Hong Kong Housing Society

In Hong Kong, many multi-storey high-density residential buildings built in the 1960s have already reached the design life and require prompt remedial works. Urban redevelopment is an effective option that can alleviate the problem, but it is becoming increasingly unacceptable as it will destruct community networks and it is not environmentally friendly. However, as aged buildings continue to pile up at an increasingly fast pace, the implementation of Mandatory Building Inspection Scheme in 2012 is an essential step to extend the building life and protect public safety. This book is a useful tool for professionals in the building and construction industry to keep up-to-date with information and practice regarding problems of ageing building, as well as opportunities in the building inspection and maintenance field.

Sr. Hon TSE Wai Chuen, Tony, BBS
LegCo Member, Architectural, Surveying and Planning Functional Constituency

Modern buildings are more than shelters. They are where we live, work, eat and play, and where we have social lives and interaction with others. Therefore, we ought to proactively manage the degradation and ageing of buildings. I would like to extend my heartfelt congratulations to Prof. Andrew Leung and his colleagues on publishing this insightful book, which provides comprehensive tools to analyse aged private buildings in Hong Kong, and to assess the demand and supply of the professional workforce related to the Mandatory Building Inspection Scheme. This book is a useful reference for policy makers, academia and professionals in the building industry.

Ir Dr Hon LO Wai Kwok, SBS, MH, JP
LegCo Member, Engineering Functional Constituency

T0325499

I am delighted to introduce this book *Mandatory Building Inspection: An Independent Study on Aged Private Buildings and Professional Workforce in Hong Kong* compiled by a research team led by Professor Andrew Y. T. Leung. This book offers a critical review on the distribution of aged private buildings in Hong Kong and evaluates the capacity of existing building professionals to meet the increasing demand arising from the full implementation of the Mandatory Building Inspection Scheme (MBIS). Professor Leung is a renowned researcher and engineer with established expertise in building construction, I firmly believe that this book will serve as an excellent reference to policy makers and building professionals with the view of tackling the problems of structural degradation and ageing buildings in Hong Kong.

Professor SHEN Qi Ping, Geoffrey
Associate Dean and Chair Professor
Department of Building and Real Estate, Hong Kong Polytechnic University

The Mandatory Building Inspection Scheme (MBIS) was fully implemented in June 2012 to ensure building safety and timely maintenance of Hong Kong's aged buildings in the wake of the 2010 Ma Tau Wai Road building collapse tragedy. The effectiveness of the MBIS not only depends on the participation of property owners but also whether there are adequate Registered Inspectors (RIs) to expertly handle every aspect of the building works. In this book, Professor Andrew Leung Yee-tak and his team examine the professional manpower needed to carry out the scheme, which I believe can serve as a resource for policy makers and building professionals seeking to better implement the MBIS and formulate future strategies to tackle the problem of building dilapidation.

Sr. CHAN Jor Kin, Kenneth
Past President, Hong Kong Institute of Surveyors
and Hong Kong Institute of Facility Management
Past Chairman 2001–2002, Surveyors Registration Board

Professor Andrew Leung's *Mandatory Building Inspection: An Independent Study on Aged Private Buildings and Professional Workforce in Hong Kong* is a must-read for all in the building maintenance business. It acquaints readers with a substantial insight of the concerned legislation and presents a comprehensive projection of the demand for building inspection professionals based on statistical analyses of the existing market as well as positive and negative incentives for qualified professionals to register as inspectors. While the content is necessarily informative and number-heavy, Prof. Leung leads his readers swiftly through the fog of digits with an articulated style of writing.

Ir WONG Tin Cheung, Conrad, BBS, JP
Chairman, Hong Kong Green Building Council

The Mandatory Building Inspection Scheme (MBIS) was implemented by the Hong Kong Buildings Department in 2012 to ensure that all ageing buildings are properly maintained so they will not turn into safety threats to the public. For the Scheme to be successful, it is necessary to have a sufficient number of Registered Inspectors who can offer professional guidance to the inspection, maintenance and repair of buildings. This book summarises the findings of a pioneering study on the statistics of aged buildings in Hong Kong and the professional work force available. With this information, a work force demand model is developed and recommendations for the effective delivery of MBIS are provided. The book is a timely contribution on a very important topic. I recommend it strongly to policy makers and practising engineers as well as students in civil engineering, building and construction programmes.

Professor LEUNG Kin Ying, Christopher
Professor, Department of Civil and Environmental Engineering
School of Engineering, Hong Kong University of Science and Technology

Mandatory Building Inspection

Mandatory Building Inspection

An Independent Study on Aged Private Buildings
and Professional Workforce in Hong Kong

Andrew Y. T. LEUNG
Michael C .P. SING
Ken H. C. CHAN

CITY UNIVERSITY OF
HONG KONG PRESS
香港城市大學出版社

ISBN: 978-962-937-239-2

Published by
    City University of Hong Kong Press
    Tat Chee Avenue
    Kowloon, Hong Kong
    Website: www.cityu.edu.hk/upress
    E-mail: upress@cityu.edu.hk

Printed in Hong Kong

Table of Contents

Detailed Chapter Contents

Preface

Along with the population and building booms of Hong Kong in the 1960s, the problem of building dilapidation was being exacerbated. Buildings in state of dilapidation will not only pose problem in terms of structural failure, hygiene, fire safety but also bring along hazards like object falling from height which would cause injury to innocent passer-by. On 29 January 2010, a five-storey residential block in 45J Ma Tau Wai Road collapsed and claimed lives of 4 residents. To the Hong Kong Government and citizens alike, this tragedy not only tolled the death bell for realising the potential danger of aged and dilapidated buildings, but also catalysed the awareness of the importance of keeping the condition of building in proper manner and the necessity of implementing mandatory building inspection and rectification scheme in Hong Kong. As at 31 December 2014, there were 43,163 numbers of private domestic and non-domestic buildings in Hong Kong among which 23,797 were over 3 storeys in height. Out of these 23,797 buildings, 15,581 buildings, equivalent to approximately 65% of the total stock, were with age 30 or above. Without proper maintenance, these buildings are no difference to bombs and landmines in the city.

To arrest the long-standing problem of building dilapidation, the Buildings Department commenced the registration of Registered Inspectors (RI) for the Mandatory Building Inspection Scheme (MBIS) on 30 December 2011. They are building professionals coming from Architects, Surveyors and Engineers. The MBIS has been fully implemented from 30 June 2012. Under the MBIS, 2000 number of buildings aged above 30 would be selected annually and the owners are required to employ the Registered Inspector for carrying out the inspection and rectification works. This book provides a critical examination on the current private building stock and existing professional workforce available for implementing the MBIS. This information can be used to project the future private building stocks that would be falling into the MBIS and thence demand of professional workforce to implement the MBIS. Therefore, this book is not only useful to policy makers, building professionals and property owners, but also to researchers and students in this field.

Chapter 1 gives us an overview on the history and latest development of building control in Hong Kong. Chapter 2 provides the statistics and analysis on the existing stock and distribution of private buildings and population density among the 18 Districts of Hong Kong. Chapter 3 reviews the historical development of building inspection scheme in Hong Kong and the scope of works under the MBIS. Chapters 4 and 5 analyse the existing stock of private buildings and provide projection of future building stock in the coming 10 years. Chapter 6 provides a review on the requirement and administrative procedures for building professionals registering as Registered Inspectors. Chapter 7 evaluates the existing supply pool of building professional workforce available for the MBIS. Chapter 8 presents the findings from the web-based questionnaire administrated to building professional and provides a projection on the demand and supply of registered inspectors in coming 10 years. Last but not least, Chapter 9 offers recommendations on how to implement and deliver the MBIS effectively.

Andrew Y. T. LEUNG
Michael C. P. SING
Ken H. C. CHAN

Acknowledgements

The Research Team would like to thank the following parties for their support and advice to this research study.

- Hong Kong Institute of Architects (HKIA)

- Hong Kong Institute of Surveyors (HKIS)

- Hong Kong Institution of Engineers (HKIE)

About the Authors

Andrew Y. T. LEUNG

MSc, PhD, DSc Aston, CEng, FRAeS, FRICS, MIStructE, MHKIE

Professor Andrew Leung taught in the University of Hong Kong and Manchester University and is now associating with City University of Hong Kong. He is Honorary/ Guest Professor of 15 universities internationally. He published more than 1,050 pieces of work in Built Environment, including 12 monographs and 500 journal papers of which 450 are cited in Science Citation Index. He is the President of Asian Institute of Intelligent Buildings, Chairman of the Chinese Green Building Council (Hong Kong), a member of the Royal Institute of Chartered Surveyors HK Board, the Join Structural Division Committee HKIE & IStructE, a Special Fellow of CIOB and HK BEAM Faculty. He is also a member of the HK BEAM Society Innovation and Addition group and of the HKGBC International and Mainland group.

Contact: andrew.leung@cityu.edu.hk

Michael C. P. SING

BSc (1st Hons), MEng, PhD (Curtin), MRICS, MAIBS, MASCE, MAIIB, ICIOB

Dr. Michael Sing completed his Bachelor Degree in Building Surveying and Master in Building Engineering in 2004. He is a Chartered Building Surveyor, and has over eight years of experience working in the field of building surveying and project management. After obtaining his PhD in Construction Management and Engineering, he served as a Senior Lecturer at Curtin University, Australia. He has published more than 50 articles in prestigious international journals and conference proceedings. His research interests involve: project performance evaluation, modeling and simulation, building control and maintenance management. Currently, he is a Visiting Assistant Professor in the Division of Building Science and Technology, City University of Hong Kong.

Contact: mcpsing@outlook.com

Ken H. C. CHAN

B App Sc, MSc, EngD, FHKIS, FRICS, FCIOB, MHKIE, MHKICM

Dr. Ken Chan is a Professional Quantity Surveyor, Building Engineer and Construction Manager. He has been involved in a number of iconic projects in Hong Kong, Macau and Dubai. In addition to being a multi-faceted professional practitioner in the construction field, he has also served as a Panel Assessor for both the Hong Kong Institute of Surveyors and the Royal Institution of Chartered Surveyors and a Council Member of the Chartered Institute of Building (Hong Kong). He has been a Tutor/Lecturer at the Curtin University of Australia and Heriot-Watt University of England, and is currently a Visiting Lecturer at the Hong Kong Polytechnic University. He received his Engineering Doctorate Degree (EngD) in Civil and Architectural Engineering in the City University of Hong Kong.

Contact: kenchan.polyuvl@gmail.com

List of Illustrations

Tables

Figures

Abbreviations

AP	Authorised Person
AP(A)	Authorised Person (Architect)
AP(E)	Authorised Person (Engineer)
AP(S)	Authorised Person (Surveyor)
ARB	Architects Registration Board
BA	Building Authority
BD	Buildings Department
BO	Buildings Ordinance
CMBS	Co-ordinated Maintenance of Buildings Scheme
ERB	Engineers Registration Board
HKHS	Hong Kong Housing Society
HKIA	Hong Kong Institute of Architects
HKIE	Hong Kong Institution of Engineers
HKIS	Hong Kong Institute of Surveyors
IRC	Inspectors Registration Committee
MBIS	Mandatory Building Inspection Scheme
MWIS	Mandatory Window Inspection Scheme
OBB	Operation Building Bright
PNAP APP	Practice Notes for Authorised Persons, Registered Structural Engineers and Registered Geotechnical Engineers
PNL	Pre-notification letter
RA	Registered Architect
RC	Registered Contractor
RGBC	Registered General Building Contractor
RI	Registered Inspector
RI(A)	Registered Inspector (Architects)
RI(E)	Registered Inspector (Engineers)
RI(S)	Registered Inspector (Surveyors)
RMWC	Registered Minor Works Contractor

RPE	Registered Professional Engineer
RPS	Registered Professional Surveyor
RSE	Registered Structural Engineer
SRB	Surveyors Registration Board
UBW	Unauthorised Buildings Works
URA	Urban Renewal Authority
VBAS	Voluntary Building Assessment Scheme

CHAPTER 1

Introduction

1.1 Background

Hong Kong is a tiny city with a total land area of 1,104 km², making it one of the most densely populated places in the world. It was also among the five densest urban areas in the world in 2014. Up to September 2014, the population of Hong Kong was 7,234,800, more than a double of that in 1961 (Census and Statistics Department, 2014).

The number of domestic buildings boomed with the rapid population growth and migration rate of Hong Kong between 1961 and 1990. A steady growth was observed in line with the constant demand of domestic buildings. In order to meet the accommodation needs, society was increasingly concerned about the quantity over quality of the buildings. As of 31 December 2014, there were 43,163 private buildings (excluding New Territories Exempted Houses) in Hong Kong among which 23,797 were non-domestic buildings and domestic buildings over 3 storeys in height. There were 15,581 private buildings, which aged 30 or above. The lack of regulation on governing maintenance work of aged private buildings also posed severe building dilapidation problems. There were a total of 143 accidents due to unsafe building structure, causing 101 deaths and 435 injuries during year 1990 and 2001 (CIRC, 2001).

According to the Buildings Department (2013a), the number of accidents related to dilapidated buildings drastically increased to 4,859, resulting in 63 deaths and 602 injuries between year 2002 and 2012. In year 2008 and 2010, the Development Bureau of the Hong Kong Government conducted an extensive condition survey on buildings aged 30 years or above, and revealed that more than 20% of those buildings were in dilapidated condition of various degrees (Development Bureau, 2011a). In order to maintain and safeguard better building safety, the Hong Kong government and citizen were increasingly aware of the necessity of establishing a systematic and mandatory building inspection and maintenance scheme to deal with the aggravated building dilapidation problems in Hong Kong.

1.2 Review of Government Policy on Building Dilapidation Problem

In the past 20 years, the Hong Kong Buildings Department (BD) is the unique authority for promoting proper repair and maintenance of aged buildings through enforcement of the Buildings Ordinance (BO). They have teams of building professionals such as Building Surveyors and Structural Engineers to deal with daily complaints from the public on the safety of existing buildings, for example, concrete spalling at the external wall. They also took the rigorous enforcement actions, for example, issuing the statutory repairs orders/notices to registered owners and requesting them to rectify the building defects within a period. To a certain degree, the existing enforcement policy is too passive and does not address the root cause of building dilapidation problems—the lack of proper maintenance works to aged buildings. Starting from November 2000, a one-off policy called Co-ordinated Maintenance of Buildings Scheme (CMBS) was launched by the Buildings Department, in association with six other government departments, including the Home Affairs Department, the Fire Services Department, etc. to assist the registered owners in pursuing a comprehensive building management and maintenance programme. Various government departments joined together and conducted a survey of the target buildings and determined the scope and nature of improvement work required. A total of 150 target buildings had been selected annually for the CMBS scheme. However, the operation was ceased in 2010 whereas another operation called "Operation Building Bright" (OBB) was implemented. It was a one-off HK$2.5 billion joint operation with the Hong Kong Housing Society (HKHS) and Hong Kong Urban Renewal Authority (URA). It provided subsidies and one-stop technical assistance to the registered owners of about 1,000 target buildings aged 30 years or above to carry out inspection and repair works. The grant was first used on repair work in common area relating to the improvement of building structural safety and sanitary facilities. However, the above policies were also too passive and not used to encourage the owners of the buildings to inspect their buildings regularly and maintain the safety condition of the buildings. To erase the building dilapidation problem in Hong Kong, owners has the undisputed responsibility to inspect and maintain their buildings in a proper condition.

1.3 The History of the Mandatory Building Inspection Scheme (MBIS)

The demand of "mandatory building inspection" could be traced back to 1985 when the Unauthorised Building Advisory Committee proposed a mandatory building inspection certification scheme. However, the scheme was dropped because of adverse public response due to the prevailing weak building care culture of owners. To deal with the long-standing building dilapidation problems, the Government had conducted two more rounds of public consultations in 2003 and 2005. The discussion includes (a) how the owners discharge their legal responsibility for inspecting and repairing the buildings, (b) any assistance available for the owners, (c) how easy the proposal scheme can be complied and (d) the role of government in the mandatory building inspection scheme. In 2007, the Government announced that it would legislate to implement the MBIS (Development Bureau, 2010a).

On 30 June 2011, the MBIS was introduced with the enactment of relevant amendments to the Buildings Ordinance through the Buildings (Amendment) Ordinance 2011 and the subsidiary legislation including the Building (Inspection and Repair) Regulation. On 30 December 2011, the Buildings Department commenced the registration for Registered

Inspectors (RIs) who are the person to carry out the inspection and supervise the repair works found necessary for the common parts, external wall and projection or signboard of the buildings. Finally, the MBIS was commenced on 30 June 2012.

1.4 The Mandatory Building Inspection Scheme

Under the policy of MBIS, owners of private buildings aged 30 years or above (except domestic buildings not exceeding 3 storeys) are required to carry out inspection (and, if necessary, repair works) of the common parts, external walls and projections or signboards of the buildings once every 10 years. The Buildings Department will select 2,000 target buildings per year for the implementation of the MBIS. Development Bureau (2010a, 2010b, 2010c) advised that the selection of target buildings would be based on a wide variety of factors, including (a) building age, (b) condition, (c) repair records and (d) locations. In addition, the registered owners need to employ an Registered Inspector (RI) to inspect the buildings and supervise related rectification works. An RI shall be a person who is registered on the Inspectors' Registry kept by the Building Authority (BA) and that he/she is to be appointed to carry out the prescribed building inspection and supervision of the prescribed building repair works. An RI could be an Authorised Person (AP), Registered Structural Engineer (RSE) or registered building professional possessing relevant work experience in the field of building construction, repair and maintenance and whose name is on the Inspectors' Registry (Buildings Department, 2012a, 2012b).

As of 22 February 2015, the total workforce of building professionals in relevant professional institutions—namely the Hong Kong Institute of Architects (HKIA), Hong Kong Institution of Engineers (HKIE) (disciplines of Building, Structural, Civil, Building Services and Materials) and Hong Kong Institute of Surveyors (HKIS) (divisions of Building Surveying and Quantity Surveying)—was 17,600. Among the 17,600 building professionals, 9,385 are registered as Registered Architects (RAs), Registered Professional Engineers (RPEs) (relevant disciplines) and Registered Professional Surveyors (RPSs) (relevant divisions). To qualify as an RI, the above building professionals need to pass a professional interview (except those nominated by their respective registration boards). The Government and the three professional institutions are considered to be overly optimistic that there should be enough building professionals for the registration of RIs in light of the size of the pool or number of corporate members in their respective institutions and the enthusiastic participation and responses of members during the course of discussion of the MBIS. As of 31 December 2014, there were 23,797 non-domestic buildings and domestic buildings over 3 storeys in height. Among them, there are 15,581 buildings aged 30 years or above failing under the MBIS. The problem of aged private buildings shall become increasingly serious in the coming decade. By year 2024, there will be 20,294 private buildings aged 30 years or above. Given the significant impacts on private building owners and professional workforce as a result of the implementation of the MBIS, it is important to examine not only the number and trend of aged buildings in the city, but also the number of potential eligible RIs available in the workforce market and to assess the future demand of the RIs.

In this book, a full review on the existing private building stock in Hong Kong was conducted using the information provided by the Home Affairs Department and extracted from the *Names of Buildings* published by the Rating and Valuation Department. A consolidated database on the private buildings was set up to reveal the real picture of the aging trend of the private building stock in Hong Kong. The number and size (in terms of number of building units) of private buildings per year group in the territory of Hong Kong and across the 18 Districts were captured to reflect the number of buildings falling under the MBIS in the coming decade.

To formulate the professional workforce planning model for the MBIS, desktop study and questionnaire survey were used to collect the data from the industry and professional institutions. A desktop study including identification and elimination of multiple memberships across architects, engineers and surveyors was conducted to examine the number of "potential" RIs coming from the relevant disciplines and divisions of the professional institutions including the Hong Kong Institute of Architects, Hong Kong Institution of Engineers and Hong Kong Institute of Surveyors. Basing upon the data, the RI registration and failure rates could be analysed and the future supply of RIs could then be portrayed. With the support of the building professional institutions, a web-based questionnaire survey was also administered to their professional corporate members via email with the following steps:

Stage 1: To assess the potential number of building professionals who would register themselves as RIs under the MBIS. Questions on reasons of registering/not registering as RIs, and participating/not-participating RIs services etc. were included.

Stage 2: To collect the workforce data related to the implementation of the MBIS. Respondents were asked to estimate the time (in terms of hours) required to carry out the inspection and repair works as stipulated in the scheme. Combining the number of aged buildings and the collected man-day of RI per building, the workforce model on projecting the demand and supply of RIs was then formulated.

The above findings could serve as a probe for the policy makers and relevant training authorities to visualise the workforce demand under the MBIS and ensure having sufficient professional workforce for the implementation of the MBIS.

References

Buildings Department (2012a). *Mandatory Building Inspection Scheme (Pamphlet on MBIS)*. Hong Kong: Buildings Department of HKSAR.

Buildings Department (2012b). *Mandatory Window Inspection Scheme (Pamphlet on MWIS)*. Hong Kong: Buildings Department of HKSAR.

Buildings Department (2013a). *Target Buildings under Mandatory Building Inspection Scheme (MBIS) and Mandatory Window Inspection Scheme (MWIS) with statutory MBIS/ MWIS Notices served by the Buildings Department*, Buildings Department of HKSAR. Available at https://mwer.bd.gov.hk/MBIS/MBISSearch.do?method=SearchMBIS (Accessed on 6 March 2013)

Census and Statistics Department (2014). *Hong Kong in Figures, 2014 Edition*. Hong Kong: Census and Statistics Department of HKSAR.

Construction Industry Review Committee (CIRC) (2001). *Construct for Excellence—Report of the Construction Industry Review Committee*. HKSAR.

Development Bureau (2010a). Government to legislate for mandatory building and window inspection schemes (with videos) (21 January 2010) [Press release]. Hong Kong: Development Bureau of HKSAR.

Development Bureau (2010b). *Legislative Council Brief (21 January 2010)—Building Ordinance (Chapter 123) and Buildings (Amendment) Bill 2010*. Hong Kong: Development Bureau of HKSAR.

Development Bureau (2010c). *Legislative Council Panel on Development (for Discussion on 25 May 2010, Review of the Urban Renewal Strategy—Stage 3 Public Engagement)*. Hong Kong: Development Bureau of HKSAR.

Development Bureau (2011a). *Examination of Estimates of Expenditure 2011–12* (Reply Serial No. DEVB(PL)(077), Question Serial No. 1046, 16 March 2011).Hong Kong: Development Bureau of HKSAR.

CHAPTER 2

Statistics of Private Buildings in Hong Kong

2.1 Introduction

This chapter focuses on the population growth rate in Hong Kong, habitation of residents in public and private buildings and population density across the 18 Districts. The existing stock and distribution of private buildings is also discussed.

2.2 An Overview of the Aged Buildings in Hong Kong

On 29 January 2010, a five-storey residential block at 45J Ma Tau Wai Road collapsed and claimed the lives of 4 residents (Buildings Department, 2010a, 2010b). According to the Buildings Department (2010b), the collapse was likely to be triggered by the disturbance of columns due to structural alteration works on the ground floor. The collapsed building was more than 50 years old. Before the incident, the Buildings Department had inspected its structure and served repair orders, but unfortunately these orders had never been compiled.

To both the Hong Kong Government and citizens, the tragedy not only tolled the death bell for them to realise the potential danger of aged and dilapidated buildings, but also catalysed the rise in awareness of the importance of proper building maintenance and necessity of imminent mandatory building inspection in Hong Kong. Figure 2.1 is the snapshots of the collapsed building at 45J Ma Tau Wai Road on 29 January 2010.

As one of the most populous places in the world, Hong Kong is home to 7,234,800 residents—more than a double of its population in the 1960s—as of December 2014, with a limited land supply. (Census and Statistics Department, 2014). Figure 2.2 outlines the population and average annual growth rate between 1961 and 2014 (Census and Statistics Department, 2014). The rapid population growth is one of the key factors accounting for the booming number of domestic building units from 1961 to 1981. The distribution on population by type of housing for the years of 2004, 2009 and 2014 is shown in Table 2.1.

Table 2.1
Distribution of Population by Type of Housing

Type of Housing	2004	2009	2014*
Public rental housing	31.10%	29.30%	29.30%
Subsidised home ownership housing	19%	18.10%	16.50%
Private permanent housing	49.10%	52%	53.70%
Temporary Housing	0.80%	0.60%	0.50%

* Provisional Figure as of 31 December 2014

Source: Census and Statistics Department, 2012; Hong Kong Housing Authority, 2014.

Figure 2.1
Snapshots of the Collapse of 45J Ma Tau Wai Road Building, 29 January 2010

Source: *Wen Wei Po*, 29 January 2010

Figure 2.2
Population and Average Annual Growth Rate, 1961–2014, Hong Kong

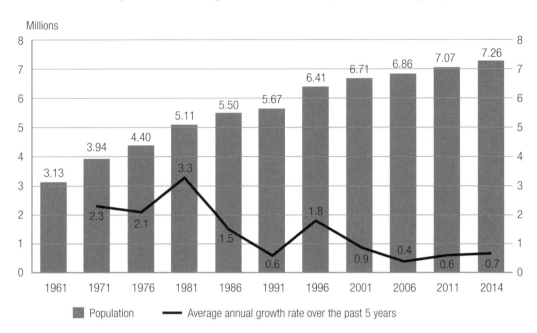

Source: Census and Statistics Department, 2014

The building dilapidation problems in Hong Kong were exacerbated mainly due to the building boom that started in the 1960s. To meet the inflated housing demands, the city is mainly concerned about the quantity of housing rather than quality in the 1960s and 1970s. Construction quality of buildings during this vintage has a far-reaching effect on the overall rate of deterioration of the city nowadays; buildings dating from the 1960s are of oddly poor quality (Chan and Tam, 2000). Most domestic buildings in Hong Kong are high-rises built with framed reinforced concrete. Buildings constructed before the 1960s are normally three to six storeys tall, according to statistics retrieved from the Rating and Valuation Department. With rapid advances in construction technology, domestic buildings of over 20 storeys have been constructed since the 1970s (Leung and Yiu, 2004).

Along with the rapid population growth rate after World War II, Leung and Yiu (2004) further illustrated the accelerated growth in the number of private housing units in Hong Kong. In the 1960s, the growth rate reached a high of 12% per annum (p.a.); this can be considered as a period of a building boom. From the 1970s to 2000s, the growth rate leveled off to about 3%, with only about 30,000 new units built per year.

According to the Hong Kong Housing Authority (2007, 2014) and the Census and Statistics Department (2012), the total growth in public and private domestic building units in Hong Kong from 1990 to 2014 was about 1,015,000 units. However, the growth in the domestic building units in the private sector was much higher than that of the public sector, with 606,000 units versus 409,000 units. Table 2.2 and Figure 2.3 show changes in domestic building units from 1990 to 2014.

Table 2.2
Change of Domestic Building Units by Ownerships from 1990 to 2014

Year	Public		Private		Total		Change in Ratio (%)	
	Units	Change (%)	Units	Change (%)	Units	Change (%)	Public	Private
1990	766,000	N/A	864,000	N/A	1,630,000	N/A	47%	53%
1997	922,000	20.37%	1,040,000	20.37%	1,962,000	20.37%	47%	53%
2004	1,085,000	17.68%	1,287,000	23.75%	2,372,000	20.90%	46%	54%
2011	1,152,000	6.18%	1,447,000	12.43%	2,599,000	9.57%	44%	56%
2014	1,175,000	2.00%	1,470,000	1.59%	2,645,000	1.77%	44%	56%
Movement 1990–2014	409,000		606,000		1,015,000			

Source: Hong Kong Housing Authority, 2007, 2014; Census and Statistics Department, 2012

Figure 2.3
Growth of Domestic Units in the Period between 1990 and 2014

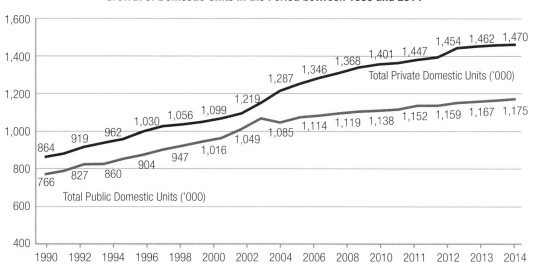

Source: Hong Kong Housing Authority, 2007, 2014; Census and Statistics Department, 2012

Figure 2.4
Proportion of Private and Public Domestic Units in 1990 and 2014 (in thousands)

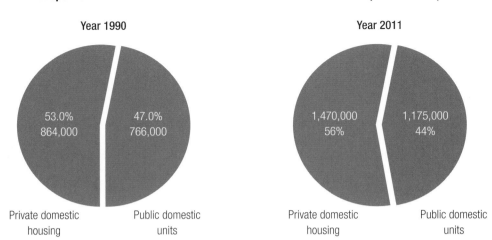

Figure 2.4 further explicates the change in proportion between private and public domestic building units from 1990 to 2014. This explains the change in government policy in the 1990s to promote the construction of private domestic housing.

As a large percentage of domestic units in Hong Kong is aging, there is an urgent call for systematic building safety and management. Legislative Council (2012b) elicited that "building safety is a complex and multi-faceted issue. If not addressed properly, the problem will become more serious". As early as 1997, the Hon Jasper Tsang Yok-sing (President of the Legislative Council from year 2008), raised concerns regarding building dilapidation in Hong Kong that "many of the private buildings completed in the 1960s or before are now in a dilapidated state. Apart from the natural cause of aging, there are a number of special attributing factors. Prior to the 1960s, there was no provision regarding safety and hygiene standards in building designs. Hence the facilities of these old buildings cannot meet the need of modern society". (Tsang, 1997, p. 129).

In February 2000, the Government set up a Task Force on Building Safety and Preventive Maintenance (the "Task Force") under the Secretary for Planning and Lands Bureau. The CIRC (2001) stated a total of 143 accidents were found related to unsafe building structure, causing 101 deaths and 435 injuries during the period from 1990 to 2001. According to the Buildings Department (2013a), the number of accidents surged to 4,859, resulting in 63 deaths and 602 injuries from 2002 till 5 December 2012.

The problems of dilapidation are especially serious for poorly managed and maintained buildings aged 20 years or above. The general awareness of owners on their obligations to maintain their properties is extremely low in Hong Kong (Ho, 1993; Yiu et al., 2007). Multiple ownership is very common in aged high-rise buildings in Hong Kong; conflicts commonly occur due to the divergent interests of different property owners in one building (Yau, 2012). Consequently, many of these are not properly managed and repair of common areas such as external wall tiles are carried out on a reactive basis.

Buildings in a state of dilapidation are potential bombs and landmines in cities, and pose problems not only in terms of structure, hygiene and fire safety, but may also be the cause of hazards like falling object from height that would lead to serious injury to innocent passers-by (Lau, 2003; Yiu et al., 2004; Sing et al., 2014). Leung and Yiu (2004) warned that a new spate of building-related accidents seems to be approaching and will create greater havoc, due to rapid high-rise and high-density development since 1955. Realising the potentially hazardous problems of building dilapidation in Hong Kong, the Government has injected substantial effort and resources into fostering a proper building maintenance culture. It was of utmost importance to finally observe a gradual change in owners' mindsets about the importance of proper building care (Development Bureau, 2007a). The Government (Development Bureau, 2010a) further advocated the slogan "Prevention is better than cure". If the building can be regularly inspected, remedial work can be carried out at the early stage to rectify the building dilapidation problem and accidents can be avoided.

Prompted by the tragedy of collapse at 45J Mau Tau Wai Road on 29 January 2010, Dr. Hon Raymond Ho Chung-tai highlighted during the Legislative Council Meeting on 4 February 2010 that "The problem is very complicated, covering very extensive areas and many different levels, urban renewal and old building maintenance is a big, old and difficult problem. This problem has told us that the problem is more urgent and complicated than the public has expected and understood" (Legislative Council, 2010). The Development Bureau had conducted a building condition survey in 2008 to 2010 for 7,000 buildings aged 30 years of above, revealing that 20% of these buildings were dilapidated to various degrees (Development Bureau, 2010d).

Sing et al. (2015) elicited that there is growing trend on the number of old buildings over the years. The results of poor and/or lack of proper building maintenance are dilapidation of buildings, and thus affecting building safety, even the occurrence of accidents which cause fatalities and injuries to occupants and the general public. Concrete spalling, unauthorised signboards, falling windows, and illegal alterations to existing building structure are urban bombs that are going to explode, and will potentially cause injuries or even fatalities (Chan et al., 2012). Lee (2010) emphasised that the dilapidated buildings have led to a poor living environment and have given rise to building safety problem. The Government should take an initiative action regarding the aspect of this problem.

2.3 Database of Private Buildings in Hong Kong

The Home Affairs Department (2014) maintains an online central database of all private buildings in Hong Kong, including residential, commercial and industrial ones. This database provides basic information to the general public about the private buildings in all 18 Districts, such as number of units and storeys, information on building management bodies, and so forth.

In addition, the Rating and Valuation Department (2014) has also published a two-volume booklet—*Names of Buildings*. Volume 1 covers Hong Kong Island and Kowloon, while Volume 2 covers the New Territories and outlying islands. This booklet—which offers a bilingual and comprehensive list of building names in the territory, such as detailed addresses and age of buildings—is available online.

Although these two sources cover "most" of the private buildings in Hong Kong, there are still several limitations. To begin with, data such as private schools, university campuses, hospitals, churches, temples are not available in the database of the Home Affairs Department. Even though both public and private buildings are in the property database created and managed by the Ratings and Valuation Department, certain types of buildings, including military establishments are excluded. In addition, some buildings are found duplicated due to the sporadic use of Chinese and/or English names of certain buildings. Moreover, the "year built" of a significant number of buildings are not shown. These include buildings exempted from rate assessment under s.36 of Cap 116 Rating Ordinance. In order to obtain authentic and trustworthy statistics of private buildings in Hong Kong, these online resources have been studied extensively to set up a comprehensive and consolidated database for the private buildings in Hong Kong.

2.4 Statistics on Private Buildings in Hong Kong

Based upon the consolidated database of private buildings in Hong Kong, as of 31 December 2014, there were a total of 43,163 private buildings, excluding the New Territories Exempted Houses under Cap 121 Buildings Ordinance (Application to the New Territories) Ordinance. Among them, and as shown in Figure 2.5, 10,173 (23.6%) were located on Hong Kong Island, 9,963 in Kowloon (23.1%) and 23,027 (53.3%) in the New Territories and Islands respectively.

The numbers, rankings and distribution of private buildings in Hong Kong was analysed in the 18 Districts (see Figure 2.6) according to the political areas as shown in Table 2.3.

As of 31 December 2014, Yuen Long (10,006 Buildings, 23.2%) and Tai Po (3,601 buildings, 8.3%) were the two districts with the densest number of private buildings in New Territories and Islands. The two districts accounted for almost one third of the total stock of private buildings in Hong Kong. The least dense district was Wong Tai Sin (507 buildings, 1.2%). However, most of the buildings in the New Territories, including Yuen Long and Tai Po, were low-rise buildings and less than 3 storeys in height. In Yuen Long, for

Table 2.3
Private Buildings in Hong Kong (As at 31 December 2014)

Item	Code	District	No. of Private Buildings in Hong Kong		Ranking
			No.	%	
1	A	Central and Western	3,632	8.4%	2
2	B	Wan Chai	2,549	5,9%	7
3	C	Eastern	1,754	4.1%	10
4	D	Southern	2,238	5.2%	9
		HONG KONG	**10,173**	**23.6%**	
5	E	Yau Tsim Mong	3,166	7.3%	4
6	F	Sham Shui Po	2,310	5.4%	8
7	G	Kowloon City	2,940	6.8%	5
8	H	Wong Tai Sin	507	1.2%	18
9	J	Kwun Tong	1,040	2.4%	14
		KOWLOON	**9,963**	**23.1%**	
10	K	Tsuen Wan	1,015	2.4%	15
11	L	Tuen Mun	1,265	2.9%	13
12	M	Yuen Long	10,006	23.2%	1
13	N	North	1,443	3.3%	12
14	P	Tai Po	3,601	8.3%	3
15	Q	Sai Kung	2,854	6.6%	6
16	R	Sha Tin	1,478	3.4%	11
17	S	Kwai Tsing	667	1.5%	17
18	T	Islands	698	1.6%	16
NEW TERRITORIES AND ISLANDS			**23,027**	**53.3%**	
TOTAL			**43,163**	**100.0%**	

Source: District Council, 2014; Home Affairs Department, 2014; Rating and Valuation Department, 2014

instance, there were 5,024 and 980 low-rise houses in the Fairview Park and Palm Springs respectively. In Tai Po, there were also 1,135 low-rise houses in Hong Lok Villa.

It is essential to note that the number of private buildings in the District Council districts may not accurately represent the dense population of Hong Kong. According to the Census and Statistics Department (2012), the densest populated district in 2011 was

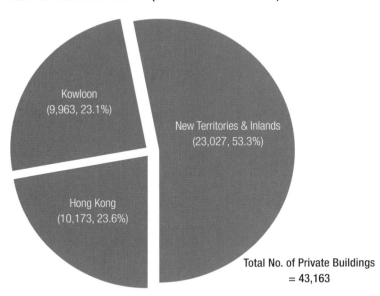

Figure 2.5
Distributions of Private Buildings in Hong Kong Island, Kowloon,
New Territories and Islands (As at 31 December 2014)

Source: Home Affairs Department, 2014; Rating and Valuation Department, 2014

Kwun Tong whereas Yuen Tong and Tai Po were ranked no. 14 and no. 17 respectively only (Table 2.4).

Yiu (2007) stated that the population density in Hong Kong in 2005 was 6,420 persons/km^2, and reaching 50,910 persons/km^2 in the most densely populated district. According to latest population census conducted in 2011, the population density was 6,552 persons/km^2. However, the population density in Kwun Tong was as high as 55,204 persons/km^2. Therefore, from 2005 to 2011, the population density in Hong Kong was an increase of 132 persons/km^2 or at a rate of 2.1%. Table 2.4 provides the population density of Hong Kong by district in 2011.

2.5 Conclusions

This chapter provides an overview of the population growth rates in Hong Kong from the 1960s. The shift in habitation of Hong Kong people from public buildings to private buildings evidenced the change in government policy towards promoting the construction of housing units by the private sector. As a large stock of private buildings ages, building safety and maintenance have attracted lots of concern from the society. The high injury

Table 2.4
Population Density (2011) by Districts of Hong Kong

Item	Code	District	Population Density 2011 (No. of Persons/km^2)[#1]	Ranking
1	A	Central and Western	20,057	3
2	B	Wan Chai	15,477	7
3	C	Eastern	31,686	6
4	D	Southern	7,173	11
		HONG KONG	**15,924**	
5	E	Yau Tsim Mong	44,045	3
6	F	Sham Shui Po	40,690	4
7	G	Kowloon City	37,660	5
8	H	Wong Tai Sin	45,181	2
9	J	Kwun Tong	55,204	1
		KOWLOON	**44,917**	
10	K	Tsuen Wan	4,918	13
11	L	Tuen Mun	5,882	12
12	M	Yuen Long	4,178	14
13	N	North	2,228	16
14	P	Tai Po	2,181	17
15	Q	Sai Kung	3,368	15
16	R	Sha Tin	9,173	10
17	S	Kwai Tsing	21,901	7
18	T	Islands	807	18
		NEW TERRITORIES AND ISLANDS	**3,876**	
		AVERAGE	**6,552**	

Note [#1] : Population density 2011– No. of persons per km^2 per District of Hong Kong in 2011
Source: Census and Statistics Department, 2014

and fatality figures resulted from aged building problem evidenced the weight of building dilapidation problems in Hong Kong. These problems are mainly the result of rapid development in high-rise and high-density buildings. The tragedy at 45J Ma Tau Wai Road was a death bell tolled at the Government and the general public, reminding them of the importance of having proper building maintenance and the need for a mandatory building inspection scheme regulated by the Government.

Figure 2.6
18 Districts of Hong Kong

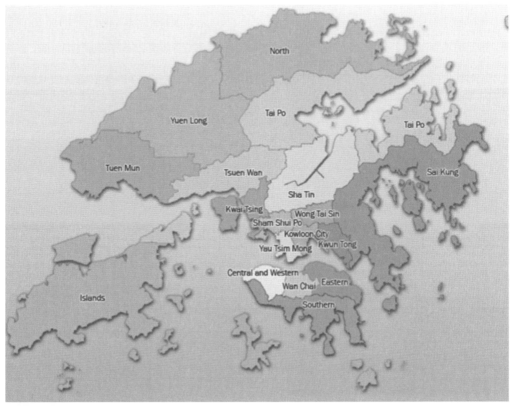

Source: District Council, 2014

As of 31 December 2014, the total number of private buildings in Hong Kong was 43,163. Among these, 10,173 (23.6%) were on Hong Kong Island, 9,963 (23.1%) were in Kowloon, and 23,027 (53.3%) were in the New Territories and Islands. Analyses from latest population census revealed that the population density of Hong Kong in 2011 was 6,552 persons/km^2. This was an increase of 132 persons/km^2 or 2.1% when compared with the year 2005. Among the 18 districts, Kwun Tong is the most densely populated district with 55,204 persons/km^2.

References

Buildings Department (2010a). Buildings Department to demolish 45G and 45H Ma Tau Wai Road (23 February 2010) [Press release]. Hong Kong: Buildings Department of HKSAR.

Buildings Department (2010b). *Report on the Collapse of the Building at 45J Ma Tau Wai Road, To Kwa Wan, Kowloon—K.I.L. 8627.* Hong Kong: Buildings Department of HKSAR.

Buildings Department (2013a). *Target Buildings under Mandatory Building Inspection Scheme (MBIS) and Mandatory Window Inspection Scheme (MWIS) with statutory MBIS/MWIS Notices served by the Buildings Department.* Buildings Department of HKSAR. Available at https://mwer.bd.gov.hk/MBIS/MBISSearch.do?method=SearchMBIS (Accessed on 6 March 2013)

Census and Statistics Department (2012). *Hong Kong in Figures (2012 Edition).* Hong Kong: Census and Statistics Department of HKSAR.

Census and Statistics Department (2014). *Hong Kong in Figures (2014 Edition).* Hong Kong: Census and Statistics Department of HKSAR.

Chan, A. P. C., and Tam, C. M. (2000). Factors affecting the quality of building projects in Hong Kong. *International Journal of Quality & Reliability Measurement, 17*(4/5), 423–441.

Chan, D. W. M., Chan, A. P. C., Lo K. K., and Hung, H. T. W. (2012). A research framework for exploring the implementation of the Mandatory Building Inspection Scheme (MBIS) in Hong Kong. *Proceedings of the Third International Conference on Construction in Developing Countries (ICCIDC–III)—Advancing Civil, Architectural and Construction Engineering and Management,* 4–6 July 2012, Bangkok, Thailand, pp. 682–687, ISBN 1-884342-02-7.

Construction Industry Review Committee (CIRC) (2001). *Construct for Excellence— Report of the Construction Industry Review Committee.* HKSAR.

Development Bureau (2007a). *Building Safety Contributes to Quality City Life.* Hong Kong: Development Bureau of HKSAR.

Development Bureau (2010a). Government to legislate for mandatory building and window inspection schemes (with videos) (21 January 2010) [Press release]. Hong Kong: Development Bureau of HKSAR.

Development Bureau (2010d). *Buildings Department Beings Inspections of Old Buildings.* Hong Kong: Development Bureau of HKSAR.

District Council (2014). *District Highlights.* District Council of HKSAR. Available at http://www.districtcouncils.gov.hk/kt/english/info/highlight_01.html (Accessed on 10 December 2014)

Ho, D. C. W. (1993). Maintenance management of aging buildings in Hong Kong. *Property Management, 11*(3), 240–245.

Home Affairs Department (2014). *Database of Private Buildings in Hong Kong.* Home Affair Department of HKSAR. Available at https://bmis.buildingmgt.gov.hk/eng/index.php (Accessed on 31 December 2014).

Hong Kong Housing Authority (2007). *Housing in Figures*. Hong Kong: Hong Kong Housing Authority of HKSAR.

Hong Kong Housing Authority (2014). *Housing in Figures*. Hong Kong: Hong Kong Housing Authority of HKSAR.

Lau, K. C. (2003). Assessing the problems of implementing major improvement works in aging private residential buildings in Hong Kong (Unpublished MSc Thesis). Hong Kong: University of Hong Kong.

Lee, W. M. (2010). *Mandatory Building Inspection Scheme, Mandatory Window Inspection Scheme and The Land (Compulsory Sale for Redevelopment)(Specification of Lower Percentage) Notice-Issue 16 (02/2010)*. Hong Kong: Maurice WM Lee Solicitors.

Legislative Council (2010). Mandatory inspection and maintenance of building*s*. *Official Record of Proceedings* (3 February 2010, pp. 4953–5037). Hong Kong: Legislative Council of HKSAR.

Legislative Council (2012b). *Subcommittee on Buildings (Amendment) Ordinance 2011 (Commencement) Notice 2012, Building (Inspection and Repair) Regulation (Commencement) Notice and Building (Minor Works)(Amendment) Regulation 2011 (Commencement) Notice—LC Paper No. CB(1)2026/11-12(01)*. Hong Kong: Legislative Council of HKSAR.

Leung, A. Y. T., and Yiu, C. Y. (2004). *Building Dilapidation and Rejuvenation in Hong Kong*. Hong Kong: City University of Hong Kong Press.

Rated and Valuation Department (2014). *Hong Kong Property Review*. Hong Kong: Rated and Valuation Department of HKSAR.

Sing, C. P., Love, P. E. D., and Davis, P. (2014). Experimental study on condition assessment of reinforced concrete Structure using a dynamics response approach.*Structural Survey, 32*(2), 89–101.

Sing, C. P., Chan, H. C., Love, P. E. D., and Leung, A. Y. T. (2015). Building maintenance and repair: Determining the workforce demand and supply for a Mandatory Building-Inspection Scheme. *ASCE Journal of Performance and Constructed Facilities*, 04015014–1.

Tsang, Y. S. (1997). The building safety inspection scheme. *Official Record of Proceedings in Council Meeting* (12 November 1997, p. 129). Hong Kong: Legislative Council of HKSAR.

Yau, Y. (2012). Multi-criteria decision making for homeowners' participation in building maintenance. *Journal of Urban Planning and Development, 138*(2), 110–120.

Yiu, C. Y., Kitipornchai, S., and Sing, C. P. (2004). Review of the status of unauthorized building works in Hong Kong. *The Journal of Building Surveying, 4*(1), 22–27.

Yiu, C. Y. (2007). Building depreciation and sustainable development. *Journal of Building Appraisal, 3*(2), 97–103.

CHAPTER 3

The Mandatory Building Inspection Scheme (MBIS) of Hong Kong

3.1 Introduction

In this chapter, the background and development of the Mandatory Building Inspection Scheme (MBIS) in Hong Kong are discussed. The requirements and procedures of the enacted MBIS are also examined in details.

3.2 Background and Development

The Mandatory Building Inspection Scheme (MBIS) and Mandatory Window Inspection Scheme (MWIS) were introduced to tackle problems arising from the lack of proper care for aging buildings in Hong Kong. The Schemes were introduced with the enactment of relevant amendments to the Buildings Ordinance through the Buildings (Amendment) Ordinance 2011 in June 2011. In December 2011, the Cap 123P Building (Inspection and Repair) Regulation was newly enforced through the amendment to the existing Cap 123 Buildings Ordinance. Under the MBIS and MWIS, building professionals such as architects, engineers and surveyors registered in relevant disciplines or divisions of various registration boards may also register as Registered Inspectors. The registration process commenced on 30 December 2011, and the MBIS and MWIS scheme have been fully implemented as of 30 June 2012. (Buildings Department, 2013a; Legislative Council, 2012a)

It is worth noting that the initiation of the MBIS was not due to SARS in 2003 or the collapse at 45J Ma Tau Wai Road in 2010. In order to provide a better understanding on the development of the MBIS and MWIS, Figure 3.1 outlines the critical events and major milestones of the Scheme.

In Figure 3.1, the demand of mandatory building inspection in the Hong Kong community started back in 1985 when the Unauthorised Building Advisory Committee proposed a mandatory building inspection certification scheme. The scheme, however, was turned down because of public reaction on the affordability of the society. In 1997, the mandatory building inspection scheme was brought back to life by Mr. Tung Chee-hwa, former Chief Executive of Hong Kong, in his first Policy Address presented to the Provisional Legislative Council on 8 October 1997. Following this direction, the Urban Renewal Authority was established in 1999 and the Government consulted the public and professional bodies about the implementation of the Mandatory Building Safety Scheme. The Government further proposed to set up a HK$500 million rehabilitation fund, aiming at providing monetary subsidies to registered owners for undertaking voluntary inspection and remedial work to their buildings (Hong Kong Government, 1997).

To study the feasibility of implementing the Mandatory Building Safety Scheme, the Planning, Environment and Lands Bureau (the predecessor of Development Bureau) conducted two rounds of public consultation in 2003 and 2005 (Development Bureau,

Figure 3.1
Critical and Major Milestones of MBIS/MWIS

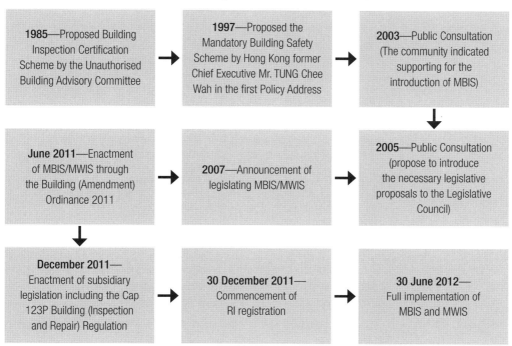

2010a). Because of the amount of accidents arising from defective buildings, the responses from the public and professional bodies were generally supportive (Yiu, 2007). In 2007, legislative procedures of reforming the existing Buildings Ordinance for facilitating the Mandatory Building Safety Scheme were initiated. The Mandatory Building Inspection Scheme (MBIS) and Mandatory Window Inspection Scheme (MWIS) were fully implemented in June 2012.

Table 3.1 summarises the timeline of the key events on the development of the MBIS and MWIS in Hong Kong.

3.3 The Mandatory Building Inspection Scheme (MBIS)

The Mandatory Building Inspection Scheme, along with the Mandatory Window Inspection Scheme, are implemented under Cap 123P Building (Inspection and Repair) Regulations for arresting the long-standing problem of neglected buildings in Hong Kong (Buildings Department, 2012a, 2012b).

- MBIS: Owners of buildings aged 30 years or above (except domestic buildings not exceeding 3 storeys) are required to carry out inspection and repair work

Table 3.1
**Summary of Major Development of the Mandatory Building Inspection Scheme (MBIS)
for Aged Private Buildings**

Date	Event
5/3/1997	Hon Christine Loh made enquiries on the details of the comprehensive building management improvement for 331 number of targeted buildings on Hong Kong Island.
12/11/1997	The implementation of the proposed Building Safety Inspection Scheme should first be targeted at the oldest or the most dangerous buildings group in Hong Kong.
20/3/1998	Funding of HK$500M is approved to provide loans for owners of domestic and domestic/commercial buildings who need financial assistance to participate in the Building Safety Inspection Scheme.
21/11/2000	Set out the idea of the Task Force on Building Safety and Preventive Maintenance on promoting timely maintenance for private buildings.
20/4/2001	LegCo Brief issued by the Government entitled "A Comprehensive Strategy for Building Safety and Timely Maintenance—Implementation Plan".
29/12/2003	A consultation paper on building management and maintenance is published for engaging the community in discussions about the appropriate approach for promoting the need for proper building management and maintenance.
21/10/2005	The Government released a public consultation paper on Mandatory Building Inspection Scheme (MBIS).
24/7/2007	Discussion was initiated on the issue of legislative procedures of Mandatory Building Inspection Scheme (MBIS) and the Mandatory Window Inspection Scheme (MWIS).
23/2/2010	The Government briefed the Legislative Council members on its measures for enhancing building safety and delivered a Summary of local press reports on building safety concerns arising from the collapse incident at Ma Tau Wai Road.
24/10/2011	The Government presented background information on the Mandatory Building Inspection Scheme and the Mandatory Window Inspection Scheme in Legislative Council.
25/10/2011	The Government briefed members on the subsidiary legislation for implementation of the Mandatory Building Inspection Scheme and the Mandatory Window Inspection Scheme.

Source: Legislative Council, 2012c

(if necessary) on the common parts, external walls and projections of buildings once every 10 years.

• MWIS: Owners of buildings aged 30 years or above (except domestic buildings not exceeding 3 storeys) are required to carry out inspection and repair work (if necessary) of all windows of buildings once every 5 years.

Based on an assessment of risk, domestic buildings not exceeding 3 storeys generally pose a lower risk to public safety. They are excluded from the MBIS and MWIS under section 30A of Cap 123 Buildings Ordinance (Development Bureau, 2010c; Development Bureau and Buildings Department, 2012)

3.3.1   Building (Inspection and Repair) Regulation

The MBIS and MWIS are enforced under Cap 123P Building (Inspection and Repair) Regulation. It empowers the Building Authority to issue statutory notices to owners requiring them to carry out prescribed inspections and repairs of their buildings and windows every 10 years and 5 years respectively. The legislation also provides for matters relating to the appointment, control and duties of Registered Inspector and Qualified Person as well as the procedural requirements for such inspections and repairs of the buildings and windows respectively. (Buildings Department, 2013a; Legislative Council, 2011d)

3.3.2   Scope of Works under MBIS

Under the MBIS, an appointed RI is to carry out prescribed inspection and supervise necessary repair work on the common parts of the buildings every 10 years (Buildings Department, 2012a). It covers the following building elements: (a) external elements and other physical elements, (b) structural elements, (c) drainage systems, (d) fire safety elements and (e) identification of any unauthorised buildings works (UBW) in the common parts of the building, on the exterior other than the common parts of the building (such as external wall, roof, podium or slope adjoining the building), or on the street which the building fronts or abuts. In case of any UBW(s) found in the building along with the MBIS, the Buildings Department will issue a separate statutory order in accordance with the prevailing enforcement policy under section 24 of Cap 123 Buildings Ordinance. Priority enforcement actions including fines and prosecution will be taken against UBWs constituting an obvious hazard or imminent danger to lives and properties, UBWs that are newly constructed, and other UBWs which are actionable items under the enforcement policy, requiring owners to demolish the same.

There were views for having longer inspection cycles to allow more time for owners to comply with the inspection and rectification programme. The Hon Michael Suen Ming-yeung, former Secretary for Housing, Planning and Lands, advised that "from the building safety angle and the need to strike a proper balance between ensuring building safety and minimising owners' burden, we are now considering the feasibility of a longer inspection cycle, say every 10 years, by reference to the date when the Buildings Department last issued the mandatory inspection notification to owners" (Development Bureau, 2007b).

Responding to concerns from some deputations regarding the consistency of building safety policy on all buildings and the question of exempting domestic buildings not exceeding three storeys in height from the MBIS and MWIS, the Development Bureau judged that category of this kind generally posed a lower risk to public safety and thus excluded it from the MBIS and MWIS (Development Bureau, 2010b, 2010c).

3.3.3   Target Buildings

The Hong Kong Institute of Surveyors reminded that "there should be some established principles and criteria for the selection of buildings to avoid arbitrary decision giving rise to appeals by building owners" (HKIS, 2012a). The target buildings would be selected based on a wide variety of factors including (a) building age, (b) building condition, (c) repair records and (d) location of the building (Development Bureau, 2010b, 2012). Furthermore, the target buildings would represent a mix of buildings in different conditions and age profiles in different districts (Buildings Department, 2012a; Development Bureau, 2010b, 2012). A selection panel, comprising representatives from various professional institutions, non-governmental organisations, property management professionals, District Councils members, and relevant Government departments, was established to solidify advice to the Buildings Department for selecting target buildings. Before serving any statutory notices for mandatory inspections, Pre-Notification Letters (PNL) would be issued to building owners advising them that their buildings were selected as target buildings and allowing them ample time to get prepared and plan ahead. Meanwhile, the Hong Kong Housing Society (HKHS) and the Urban Renewal Authority (URA) would stand ready to provide support and organise a briefing session for the owners. Details of the scheme, including the requirements of employing RI, would be introduced to the owners.

To facilitate the progress of the MBIS and MWIS, a web-based system—the Building Register for the Mandatory Building Inspection Scheme and the Mandatory Window Inspection Scheme—has been launched by the Buildings Department. The system allows owners, owners' corporations and the general public to view or make reference to those target buildings the MBIS and/or MWIS statutory notices have been served. The list is compiled with addresses and districts of the buildings being selected and the dates of the statutory notices being served. The list will be updated on a weekly basis. As of December 2014, there was more than 4,600 private buildings joined the MBIS and MWIS.

3.3.4   Procedures under MBIS

In the early stage, the Buildings Department will issue Pre-Notification Letters to owners or owners' corporations of the selected target buildings, alerting, preparing and organising them to carry out the prescribed inspection and repair. Six months after issuing the PNLs, statutory notices will be issued to the owners requiring them to carry out the prescribed inspection and repair work necessary in respect of the common parts, external walls and projections, or signboards within the specified time frame. A Registered Inspector should be appointed to carry out the prescribed inspection to ascertain whether the building has been rendered dangerous or is liable to become dangerous within the specified time frame. If the RI considers that a prescribed repair to be necessary, the owners will need to appoint a Registered Contractor (RC) to carry out the prescribed repair under the supervision of the RI. The appointed RC under the MBIS shall be an Registered General Building Contractor (RGBC) or a Registered Minor Works Contractor (RMWC) who is qualified to carry out the rectification and repair works. Upon completion of the prescribed inspection and repair, the RI has to submit an endorsed inspection report and a completion report respectively, together with a certificate in the specified form, to the Building Authority for record and audit check. The total allotted time frame from receiving the formal MBIS statutory notice to submission of the completion report is 12 months. An extra three months will be allowed for those private buildings without an owners' corporation (Buildings Department, 2013b). Figure 3.2 presents the timeline and procedures under the MBIS.

3.3.5   Unauthorised Building Works identified under MBIS

In respect of the unauthorised building works (UBWs) identified during the inspection of the RI, the Buildings Department (2012a) further advised that "RIs have to report to the BD for any UBWs identified in the common parts of the building, on the exterior other than the common parts of the building (such as external wall, roof or podium, yard or slope adjoining the building) or on the street on which the building fronts or abuts that the RI inspected".

3.3.6   Registered Inspectors

Under section 30D(1)(a) or (b) of Cap 123 Buildings Ordinance, the Registered Inspectors should be appointed to carry out the inspection and repair works as prescribed under the MBIS. The RI must have practical knowledge to deal with statutory requirements related to building design, building diagnosis, remedial methodology, supervision, and specification of maintenance works (HKIS, 2012a). An RI could be an Authorised Person, Registered Structural Engineer, or registered building professional possessing relevant work experience in the field of building construction, repair and maintenance, whose name is on the Inspectors' Registry (Buildings Department, 2011). The RIs play an important role in the implementation of the MBIS. Early registration of qualified candidates to be an RI is encouraged within the professional bodies (Development Bureau, 2011e).

To ease the possible shortage of RIs at the early stage of the MBIS, suggestions were made to include the Clerks of Works, Inspectorate, and other building practitioners. Development Bureau and Buildings Department (2012) stressed that the pool of Registered Inspectors for building inspections under the MBIS has been expanded from

Figure 3.2 Procedures under the Mandatory Building Inspection Scheme (MBIS)

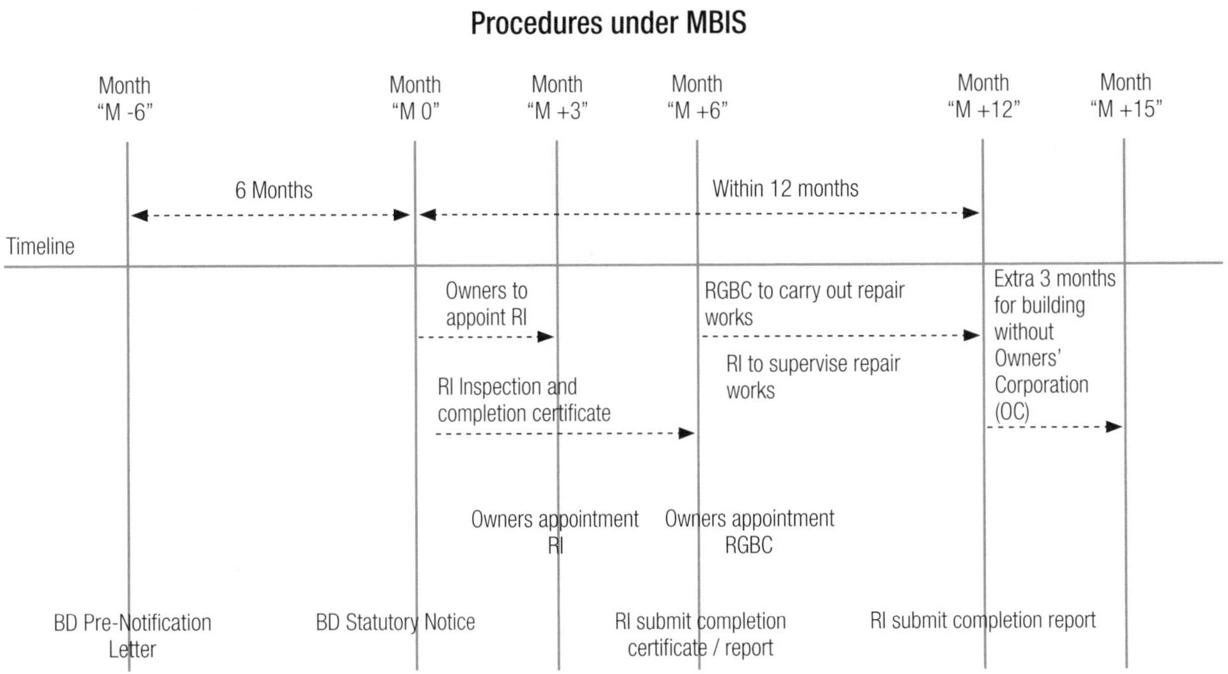

Procedures under MBIS

Timeline	Month "M -6"	Month "M 0"	Month "M +3"	Month "M +6"	Month "M +12"	Month "M +15"

6 Months

Within 12 months

Extra 3 months for building without Owners' Corporation (OC)

Owners to appoint RI

RGBC to carry out repair works

RI to supervise repair works

RI Inspection and completion certificate

Owners appointment RI

Owners appointment RGBC

BD Pre-Notification Letter

BD Statutory Notice

RI submit completion certificate / report

RI submit completion report

Explanatory Notes

1. Month M -6   BD issues pre-notification letters to the owners of the selected target buildings
2. Month M 0   BD issues statutory notices to the owners of the target buildings
3. Month M 0 to M +3   Owners to appoint Registered Inspector (RI)
4. Month M 0 to M +6   RI to carry out prescribed inspection and submit completion certificate, report and if necessary a repair proposal
5. Month M +6 to M +12   If prescribed repair is required, owners to appoint a RGBC and RI
6. Month M +6 to M +12   RGBC to carry out the prescribed repair works
7. Month M +6 to M +12   RI to supervise the prescribed repair works
8. Month M +12   RI to submitcompletion report
9. Month M +15   Extra 3 months allowed for building without Owners' Corporation (OC)
10. Annual Target   2,000 MBIS buildings and 5,800 MWIS (of which 2,000 are synchronised with MBIS buildings)

Source: Buildings Department, 2013b

APs and RSEs to RAs, RPEs, and RPSs of relevant disciplines/divisions, and that RIs have to be held personally responsible for the inspection and supervision of repair. The RIs should possess adequate professional knowledge and experience in building design, construction and maintenance, and be fully acquainted with the Buildings Ordinance; therefore, all RIs have to be "building professionals" who are recognised by their respective professional registration boards to be professionally competent.

3.3.7   Duties of Registered Inspectors

According to the Buildings Department (2013a), duties of RIs are specified in the Buildings Ordinance and its regulations and the Code of Practice which include, but are not limited to, carry out personal inspection, provide proper supervision, ensure the repair materials and their use are in compliance with the Buildings Ordinance (BO) and relevant standards, ensure the repair works are safe and will render the building safe, notify the Building Authority (BA) and owners of any emergency, contravention of the provisions of the BO or appointment of the RI and RGBC, etc., and submit relevant documents to the BA and the owners.

Under section 6.4 of the Code of Practice for the Mandatory Building Inspection Scheme and the Mandatory Window Inspection Scheme, the RI has to take the overall responsibility for the supervision of the Registered Contractor (Buildings Department, 2011). However, during the rectification and repair work stage, the RI can designate a person as his/her representative to provide supervision to the Registered Contractor. Duties of the RI whom may be assisted by his/her representative, include:

(a) conduct building inspection to ascertain the extent of defects that s/he established the nature and cause of; and

(b) supervise the rectification and repair work and deficiencies identified during the inspection and repair stages.

The basic qualification and experience requirements of RI representatives include a certificate or diploma in a construction related discipline—i.e., architecture, building studies, engineering or surveying, with a minimum of 2 years of relevant work experience.

3.3.8   Regulation of Service Providers

Under the MBIS (Buildings Department, 2012c), the Buildings Department will ensure proper regulation of service providers under the Buildings Ordinance and its subsidiary legislation, including the Building (Inspection and Repair) Regulation.

It is worth noting that despite the introduction of a new category of building professional, the Registered Inspector, the Buildings Department will not depart nor shift its roles and responsibilities for upkeeping a safe and healthy building environment for the community to the RI.

3.3.9 The Voluntary Building Assessment Scheme (VBAS)

Under the MBIS, the Hong Kong Housing Society has also launched a similar scheme— the Voluntary Building Assessment Scheme (VBAS) to give positive recognition to well-managed and properly maintained buildings (Hong Kong Housing Society, 2013). Buildings certified by the VBAS will be recognised by the Buildings Department for having fulfilled the requirements under the MBIS and MWIS (in respect of windows in the common parts only) within the respective inspection cycles. The HKHS has been receiving applications from building owners for participating in the VBAS since July 2012.

3.4 Conclusions

The underlying spirit and intrinsic value of the Schemes are to inspect the buildings on a regular basis, to identify problems at an early stage, and to carry out remedial work on time so as to avoid accidents related to building dilapidation.

The demand for "mandatory building inspection" in Hong Kong can be traced back to as early as 1985 but the scheme was dropped because of adverse public reaction. A two-round public consultation was conducted in 2003 and 2005, upon which the Government announced in 2007 that it would legislate the implementation of the MBIS and MWIS. The MBIS and MWIS were introduced with the enactment of relevant amendments to the Buildings Ordinance through the Buildings (Amendment) Ordinance 2011 in June 2011 and the subsidiary legislation including Cap 123P Building (Inspection and Repair) Regulation in December 2011. The Buildings Department commenced the registration for Registered Inspectors on 30 December 2011 and the full implementation of the MBIS and MWIS was commenced on 30 June 2012.

Under the MBIS, the Buildings Department will select 2,000 target buildings per year. Owners of buildings aged 30 years or above (except those domestic buildings not exceeding three storeys) will be required to appoint a new category of building profession, Registered Inspectors, to carry out the inspection and, if necessary, supervise the repair work of common parts, external walls and projections of the buildings once every 10 years.

References

Buildings Department (2011). *Registration of MBIS Registered Inspectors Begins, APP–7.* Buildings Department of HKSAR. Available at http://www.bd.gov.hk/english/documents/news/20111229Bae.htm (Accessed on 21 March 2012)

Buildings Department (2012a). *Mandatory Building Inspection Scheme (Pamphlet on MBIS).* Buildings Department of HKSAR, Hong Kong.

Buildings Department (2012b). *Mandatory Window Inspection Scheme (Pamphlet on MWIS).* Buildings Department of HKSAR, Hong Kong.

Buildings Department (2012c). *Inspectors' Register.* Buildings Department of HKSAR, Available at http://www.bd.gov.hk/english/inform/index_ap.html (Accessed on 22 February 2015)

Buildings Department (2013a). *Target Buildings under Mandatory Building Inspection Scheme (MBIS) and Mandatory Window Inspection Scheme (MWIS) with statutory MBIS/MWIS Notices served by the Buildings Department.* Buildings Department of HKSAR. Available at https://mwer.bd.gov.hk/MBIS/MBISSearch.do?method=SearchMBIS (Accessed on 6 March 2013)

Buildings Department (2013b). *Monthly Digest (2005–2012)—Buildings For Which Building Authority has issued Demolition Consent.* Buildings Department of HKSAR. Available at http://www.bd.gov.hk/english/documents/index_statistics.html (Accessed on 8 March 2013)

Development Bureau (2007b). *Mandatory Building Inspection—Hong Kong Housing Society's Assistance to Eligible Building Owners* (Ref. DEVB(PL-B)/68/03/21). Hong Kong: Development Bureau of HKSAR.

Development Bureau (2010a). Government to legislate for mandatory building and window inspection schemes (with videos) (21 January 2010) [Press release]. Hong Kong: Development Bureau of HKSAR.

Development Bureau (2010b). *Legislative Council Brief (21 January 2010)—Buildings Ordinance (Chapter 123) and Buildings (Amendment) Bill 2010.* Hong Kong: Development Bureau of HKSAR.

Development Bureau (2010c). Legislative Council Panel on Development for Discussion on 25 May 2010—*Review of the Urban Renewal Strategy – Stage 3 Public Engagement.* Hong Kong: Development Bureau of HKSAR.

Development Bureau (2011e). *Updated background brief on Mandatory Building Inspection Scheme and Mandatory Window Inspection Scheme* (Legislative Council Panel on Development, LC Paper No. CB(1)137/11-12(06), 25 October 2011). Hong Kong: Development Bureau of HKSAR.

Development Bureau (2012). Details of Mandatory Building Inspection Scheme being worked out (16 March 2012) [Press release]. Hong Kong: Development Bureau of HKSAR.

Development Bureau and Buildings Department (2012). *Buildings Department's Review of Enforcement Procedures and Practices for Dilapidated Buildings and Views of Independent Experts* (Legislative Council on Development, Subcommittee on Building Safety and Related Issue, CB(1)2099/11-12(01) 11 June 2012). Hong Kong: Development Bureau of HKSAR.

Hong Kong Government (1997). *Hong Kong Policy Address 1997–1998*. Hong Kong: Hong Kong Government Publisher.

Hong Kong Housing Society (2013). *Voluntary Building Assessment Scheme— Certificated Buildings*. Hong Kong Housing Society. Available at http://vbas.hkhs.com/en/certified_ buildings/statistics_on_applications_for_vbas_building_certification.php (Accessed on 1 February 2013).

Hong Kong Institute of Surveyors (2012a). Registration of registered inspector under MBIS (Building Surveying Division Chairman's Message). *Surveyor Times* (January 2012). Hong Kong: Hong Kong Institute of Surveyors.

Hong Kong Institute of Surveyors (HKIS) (2015). *Find a Member*. The Hong Kong Institute of Surveyors. Available at http://www.hkis.org.hk/en/membership_find.php (Accessed on 22 February 2015).

Legislative Council (2011d). *Legislative Council Brief, Subsidiary Legislation for Implementation of Mandatory Building Inspection and Mandatory Window Inspection Scheme* (File Ref.: DEVB (PL-CR) 2/15-08), October 2011). Hong Kong: Legislative Council of HKSAR.

Legislative Council (2012a). Mandatory building and window schemes. *Official Record of Proceedings* (13 June 2010, pp. 15016–15023). Hong Kong: Legislative Council of HKSAR.

Legislative Council (2012c). Paper for the House Committee meeting on 1 June 2012, *Report of Bills Committee on Building Legislation (Amendment) Bill 2011* (LC Paper No. CB(2)2191/11-12, 1 June 2012). Hong Kong: Legislative Council of HKSAR.

Yiu, C. Y. (2007). Building depreciation and sustainable development. *Journal of Building Appraisal, 3*(2), 97–103.

CHAPTER 4

Analyses on Private Buildings under the Mandatory Building Inspection Scheme

4.1 Introduction

This chapter provides an examination on the number, distribution and ageing rate of private buildings (non-domestic buildings and domestic buildings over 3 storeys in height) in 2014 and 2024 (referred as "private building" hereafter) in Hong Kong, Kowloon, and the New Territories and Islands that fit the selection criteria of the MBIS.

4.2 Private Buildings under the Mandatory Building Inspection Scheme

4.2.1 An Overview

The consolidated database of private buildings in Hong Kong reveals, as at 31 December 2014, out of the total 43,163 private buildings, 23,797 were private non-domestic buildings and domestic buildings over 3 storeys in height. This number represents approximately 55% of the total number of private buildings in Hong Kong and private buildings under this group will be under the selection criteria of the MBIS once their ages reach 30 and over. On the other hand, there were a total of 19,366 (45%) private domestic buildings equal or less than 3 storeys in height.

Table 4.1 shows the distributions of private buildings in Hong Kong with Group A representing non-domestic private buildings and domestic private buildings over 3 storeys in height. Group B represents private domestic buildings equal or less than 3 storeys in height. The distributions of private buildings between the two groups differ tremendously. These differences can be illustrated by comparing the number (or percentage) of buildings in Hong Kong Island, Kowloon, and the New Territories and Islands. For private buildings under Group A, out of the total 23,797, Kowloon shared the largest proportion with 9,274 (39%), Hong Kong Island at second with 7,965 (33.5%), and New Territories and Islands the least with 6,558 (27.5%).

On the other hand, the proportion in Group B is completely different. Out of the total 19,366, New Territories and Islands shared the largest proportion and had as many as 16,469 (85%) private buildings. Hong Kong Island was the second with a total number of 2,208 (11.4%). Kowloon had the least with a total number of 689 (3.6%).

The differences in number and percentage of private buildings which are grouped in (a) private non-domestic buildings, (b) domestic buildings, and (c) private non-domestic building less than 3 storeys in height are illustrated in Figure 4.1. The comparison of the number of private buildings in the 18 districts of Hong Kong is provided in Figure 4.2.

Table 4.1
Distribution of Private Buildings (Non-domestic/Domestic over 3 Storeys and Domestic Equal or Less Than 3 Storeys) in Hong Kong (As at 31 December 2014)

Item	Code	District	No. of Private Buildings in Hong Kong					
			Group A (Note #1)		Group B (Note #2)		Total	
			(a)	% (Note #3)	(b)	% (Note #4)	(c)=(a)+(b)	% (Note #5)
1	A	Central and Western	3,162	13.3%	470	2.4%	3,632	8.4%
2	B	Wan Chai	2,071	8.7%	478	2.5%	2,549	5.9%
3	C	Eastern	1,728	7.3%	26	0.1%	1,754	4.1%
4	D	Southern	1,004	4.2%	1,234	6.4%	2,238	5.2%
		HONG KONG	**7,965**	**33.5%**	**2,208**	**11.4%**	**10,173**	**23.6%**
5	E	Yau Tsim Mong	3,041	12.8%	125	0.6%	3.166	7.3%
6	F	Sham Shui Po	2,081	8.7%	229	1.2%	2,310	5.4%
7	G	Kowloon City	2,629	11%	311	1.6%	2,940	6.8%
8	H	Wong Tai Sin	497	2.1%	10	0.1%	507	1.2%
9	J	Kwun Tong	1,026	4.3%	14	0.1%	1,040	2.4%
		KOWLOON	**9,274**	**39.0%**	**689**	**3.6%**	**9,963**	**23.1%**
10	K	Tsuen Wan	865	3.6%	150	0.8%	1,015	2.4%
11	L	Tuen Mun	784	3.3%	481	2.5%	1,265	2.9%
12	M	Yuen Long	939	3.9%	9,067	46.8%	10,006	23.2%
13	N	North	710	3.0%	733	3.8%	1,443	3.3%
14	P	Tai Po	835	3.5%	2,766	14.3%	3,601	8.3%
15	Q	Sai Kung	480	2.0%	2,374	12.3%	2,854	6.6%
16	R	Sha Tin	1003	4.2%	475	2.5%	1,478	3.4%
17	S	Kwai Tsing	639	2.7%	28	0.1%	667	1.5%
18	T	Islands	303	1.3%	395	2.0%	698	1.6%
		NEW TERRITORIES AND ISLANDS	**6,558**	**27.5%**	**16,469**	**85.0%**	**23,027**	**53.3%**
		TOTAL	**23,797**		**19,366**		**43,163**	

Note #1:   Group A refers to the total of private non-domestic buildings and private domestic buildings over 3 storeys
Note #2:   Group B refers to the total of private domestic buildings equal or less than 3 storeys
Note #3:   Percentage (%) out of the total number of buildings (under Group A) in Hong Kong
Note #4:   Percentage (%) out of the total number of buildings (under Group B) in Hong Kong
Note #5:   Percentage (%) out of the total number of buildings in Hong Kong

Source: Home Affairs Department, 2014; Rating and Valuation Department, 2014

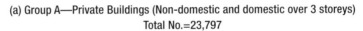

Figure 4.1
Comparison of Private Buildings in Hong Kong Island, Kowloon,
and New Territories and Islands (As at 31 December 2014)

(a) Group A—Private Buildings (Non-domestic and domestic over 3 storeys)
Total No.=23,797

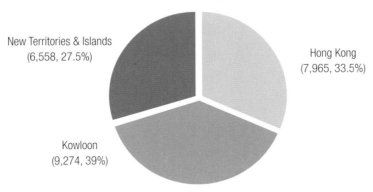

(b) Group B—Private Buildings (Domestic Equal or Less Than 3 storeys)
Total No.=19,366

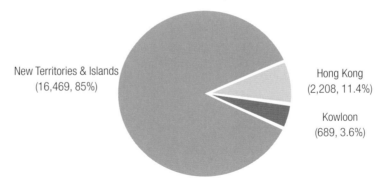

Source: Home Affairs Department, 2014; Rating and Valuation Department, 2014

4.2.2 Projected Number of Private Buildings from 2014 and 2024 under MBIS

Under the MBIS, owners of buildings aged 30 years or above (except domestic buildings not exceeding 3 storeys) are required to appoint a Registered Inspector (RI) to carry out the prescribed inspection and supervise the prescribed repair work found necessary of the common parts, external walls, and projections or signboards of the buildings once every 10 years (Buildings Department, 2012a). It is important to note that not all private buildings in Hong Kong are under the Scheme. Domestic private buildings equal or less than 3 storeys in height, New Territories Exempted Houses (NTEH), as well as all

Figure 4.2
Distribution of Private Buildings in 18 Districts
of Hong Kong (As at 31 December 2014)

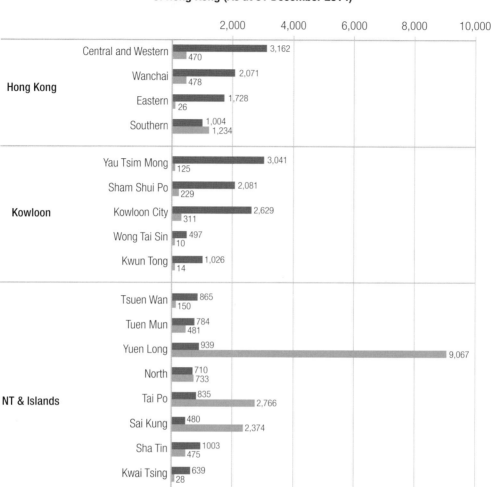

■ Total private non-domestic buildings and private domestic buildings over 3 storeys

■ Total private domestic buildings equal or less than 3 storeys

Source: Home Affairs Department, 2014; Rating and Valuation Department, 2014

government and public rental buildings under the Hong Kong Housing Authority are exempted from the MBIS. (Tsui, 2012).

Utilising the analysis on the current stock of private buildings in Chapter 2, the total number of private domestic buildings and non-domestic buildings over 3 storeys in height as at 31 December 2014 was 23,797. As shown in Table 4.2, 15,581 out of 23,797 private buildings were aged 30 or above. This figure represents 65.5% of the total stock of private

Table 4.2

Change in Numbers of Private Buildings under MBIS in 2014 and 2024

Item	District	Private Non-domestic Buildings and Domestic Buildings over 3 Storeys		2014 Buildings Aged, 30 or above				2024 Buildings Aged, 30 or above				Ageing of Private Buildings (2014 to 2024)		
		No.	Ranking	No.	Ranking	Density	Ranking	No.	Ranking	Density	Ranking	No.	Percentage (%)	Ranking
1	Central and Western	3,162	1	2,412	2	76.3%	5	2,902	1	91.8%	2	490	15.5%	13
2	Wanchai	2,071	5	1,621	5	78.3%	4	1,931	4	93.2%	1	310	15.0%	14
3	Eastern	1,728	6	1,176	6	68.1%	8	1,559	6	90.2%	6	383	22.2%	11
4	Southern	1,004	8	620	8	61.8%	9	819	8	81.6%	12	199	19.8%	12
	HONG KONG	**7,965**		**5,829**		**73.2%**		**7,211**		**90.5%**		**1,382**	**17.4%**	
5	Yau Tsim Mong	3,041	2	2,478	1	81.5%	1	2,738	2	90.0%	2	260	8.5%	18
6	Sham Shui Po	2,081	4	1,660	4	79.8%	3	1,888	5	90.7%	4	228	11.0%	16
7	Kowloon City	2,629	3	2,131	3	81.1%	2	2,408	3	91.5%	3	277	10.5%	17
8	Wong Tai Sin	497	16	340	12	68.4%	6	414	16	83.3%	11	74	24.9%	15
9	Kwun Tong	1,026	7	677	7	66.0%	7	926	7	90.3%	5	249	24.3%	10
	KOWLOON	**9,274**		**7,286**		**78.6%**		**8,374**		**90.3%**		**1,088**	**11.7%**	
10	Tsuen Wan	865	11	502	9	58.0%	10	748	9	86.5%	9	246	28.4%	7
11	Tuen Mun	784	13	254	16	32.4%	15	657	11	83.8%	10	403	51.4%	1
12	Yuen Long	939	10	362	10	38.6%	12	601	12	64.0%	17	239	25.5%	9
13	North	710	14	269	14	37.9%	13	473	15	66.6%	16	204	28.7%	6
14	Tai Po	835	12	311	13	37.2%	14	582	13	69.7%	14	271	32.5%	4
15	Sai Kung	480	17	55	18	11.5%	18	179	18	37.3%	18	124	25.8%	8
16	Sha Tin	1003	9	260	15	25.9%	17	672	10	67.0%	15	412	41.1%	3
17	Kwai Tsing	639	15	361	11	56.5%	11	555	14	86.9%	8	194	30.4%	5
18	Islands	303	18	92	17	30.4%	16	242	17	79.9%	13	150	49.5%	2
	NEW TERRITORIES AND ISLANDS	**6,558**		**2,466**		**37.6%**		**4,709**		**71.8%**		**2,243**	**34.2%**	
	TOTAL	**23,797**		**15,581**		**65.5%**		**20,294**		**85.3%**		**4,713**	**19.8%**	

Source: Home Affairs Department, 2014; Rating and Valuation Department, 2014

buildings (non-domestic buildings and domestic buildings over 3 storeys in height) falling under the scope of the MBIS. It is predicted that, in the next 10 years (i.e., in December 2024), the number of private buildings aged 30 years or above will drastically increase to 20,294, equivalent to 85.3% of the total stock of private buildings falling under the MBIS.

As of 31 December 2014, Yau Tsim Mong District ranked top with the largest number of private buildings (2,478) falling within the range of the MBIS, among the 18 districts in Hong Kong. Central and Western District ranked second (2,412). However, this position will reverse in a decade (i.e., in 2024). Central and Western District will take over the position of Yau Tsim Mong with 2,902 private buildings falling within the region of the MBIS, whereas there will only be 2,638 in Yau Tsim Mong. In 2014, the Islands with 92 number of buildings aged 30 or above ranked the last among the 18 districts, and Sai Kung (55 numbers) was the penultimate. In 2024, the numbers of private buildings falling under the MBIS in the Islands will increase to 242 and Sai Kung to 179 respectively. As such, Table 4.2 illustrates the change in numbers of private buildings falling under the MBIS in 2014 and 2024.

Forecasting from the year 2014, there will be a net increase of 4,713 private buildings falling under the MBIS in 2024. Among the three geographical regions of Hong Kong, the New Territories and Islands will have the largest increase in number, from 2,466 (37.6%) to 4,709 (71.8%), representing a net rise of 2,243 (34.2%). Hong Kong Island will rank the second, with an increase in number from 5,829 (73.2%) to 7,211 (90.5%), representing a net increase of 1,382 (17.4%). Comparing the least increase in number among the three regions of Hong Kong, i.e., Hong Kong Island, Kowloon, the New Territories and Islands, Kowloon will rank the first with an increase from 7,286 (78.6%) to 8,374 (90.3%), a net increase of 1,382 (17.4%). The change in number and proportion of aged buildings falling under the MBIS in the three regions from 2014 to 2024 is illustrated in Figure 4.3.

Figure 4.3
Comparison of Private Buildings in Hong Kong Island,
Kowloon and New Territories and Islands aged 30 or above (Years 2014 and 2024)

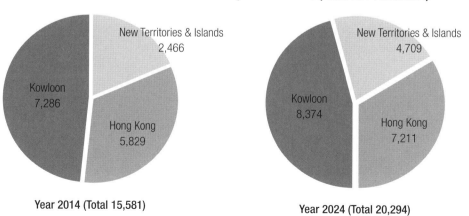

Year 2014 (Total 15,581)   Year 2024 (Total 20,294)

Note: For Non-domestic Private Buildings and Domestic Private Buildings over 3 Storeys

4.2.3 Ageing of Private Buildings in 2014 and 2024 under MBIS

As per the above analysis, in 2014, 15,581 out of the total of 23,797 private buildings (non-domestic buildings and domestic buildings over 3 storeys in height), reached the age of 30 years or above and fell within the scope of the MBIS. This figure represents an overall percentage of 65.5% of the total number of private buildings in Hong Kong. Among the 18 districts of Hong Kong, analysis reveals that the individual densities of 7 districts were higher than 65.5% the overall percentage in the territory while that of 11 districts were lower.

In 2024, the number of private buildings in Hong Kong aged 30 years or above will increase to 20,294 and then the overall percentage will rise to 85.3%. Coincidentally, among the 18 districts of Hong Kong, 7 districts will still have individual densities higher than 85.3%, while the other 11 districts will have lower densities. The rankings and densities of the 18 districts in 2014 and 2024 are reflected in Table 4.3.

Table 4.3
Rankings and Densities of Aged Private Building, 2014 and 2024
(As of 31 December 2014)

2014			2024		
Ranking	District	Density	Ranking	District	Density
	Overall Hong Kong	65.5%		Overall Hong Kong	85.3%
1	Yau Tsim Mong	81.5%	1	Wan Chai	93.2%
2	Kowloon City	81.1%	2	Central and Western	91.8%
3	Sham Shui Po	79.8%	3	Kowloon City	91.6%
4	Wan Chai	78.3%	4	Sham Shui Po	90.7%
5	Central and Western	76.3%	5	Kwun Tong	90.3%
6	Wong Tai Sin	68.4%	6	Eastern	90.2%
7	Kwun Tong	66.0%	7	Yau Tsim Mong	90.0%
8	Eastern	68.1%	8	Kwai Tsing	86.9%
9	Tsuen Wan	58.0%	9	Tsuen Wan	86.5%
10	Southern	61.8%	10	Tuen Mun	83.8%
11	Kwai Tsing	56.5%	11	Wong Tai Sin	83.3%
12	Yuen Long	38.6%	12	Yuen Long	81.6%
13	North	37.9%	13	Southern	79.9%
14	Tai Po	37.2%	14	Tai Po	69.7%
15	Tuen Mun	32.4%	15	Shatin	67.0%
16	Islands	30.4%	16	North	66.6%
17	Sha Tin	25.9%	17	Yuen Long	64.0%
18	Sai Kung	11.5%	18	Sai Kung	37.3%

More Densely District

Less Densely District

In Table 4.3, it is worth noting that, as of 31 December 2014, the first three districts with the densest aged buildings (in terms of number of buildings aged 30 years or above per total number of buildings in the corresponding district) were located in Kowloon. Yau Tsim Mong (81.5%) was the densest, Kowloon City (81.1%) ranked the second, and Sham Shui Po (79.8%) the third. However, in 2024, the ranking will change with Wanchai (93.2%) becoming the densest district in Hong Kong, Yau Tsim Mong (90%) will drop to the seventh place.

In 2014, the least dense district was Sai Kung (11.5%), and Shatin (25.9%) was the penultimate. In 2024, although the ranking of Sai Kung will remain unchanged, the percentage of aged private buildings will soar to 37.3%, more than a triple of that of 2014. For Shatin, although the ranking will notch up to the 15th, the percentage of aged private buildings will drastically soar to 67%, over 2.5 times that of 2014.

Table 4.4
Private Building Ageing Severity in Density (%), 2014 to 2024

Ageing, 2014 to 2024		
Ranking	District	Ageing %
	Overall Hong Kong	19.8%
1	Tuen Mun	51.4%
2	Islands	49.5%
3	Shatin	41.1%
4	Tai Po	32.5%
5	Kwai Tsing	30.4%
6	North	28.7%
7	Tsuen Wan	28.4%
8	Sai Kung	25.8%
9	Yuen Long	25.5%
10	Kwun Tong	24.3%
11	Eastern	22.2%
12	Southern	19.8%
13	Central and Western	15.5%
14	Wan Chai	15.0%
15	Wong Tai Sin	14.9%
16	Sham Shui Po	11.0%
17	Kowloon City	10.5%
18	Yau Tsim Mong	8.5%

Severity in Growing Density (%) of Aged Private Buildings

Analysis reveals that the earlier the district was developed, the lesser is the rate of ageing in terms of density (%) of aged private buildings. In another words, the later the development of the district, the faster the rate of ageing. Table 4.4 measures the rate of ageing in terms of density (%) of aged private buildings under the MBIS from 2014 to 2024. It is noted that the overall change in number of private buildings aged 30 years or above from 2014 to 2024 is 4,713, which is equivalent to 19.8% of the total number of private buildings (i.e., 23,797) in Hong Kong. Further analysis reveals that, in 2024, the individual rates of ageing in 12 districts will be higher than the overall rate of 21% while 6 districts will be lower.

The top three districts, in terms of rates of ageing in density (%) of aged private buildings from 2014 to 2024, will be Tuen Mun (51.4%), Islands (49.5%) and Shatin (41.1%). On the other hand, the least three districts will be Sham Shui Po (11.0%), Kowloon City (10.5%) and Yau Tsim Mong (8.5%).

4.2.4   Size Distributions of Private Buildings under MBIS

Figure 4.4 illustrates the cumulative number of private buildings (per unit groups) in Hong Kong from Pre-war period to 2014; over 90% of private buildings from Pre-war years to 1961 were small in size (building units ≤50 numbers). The size of private buildings increased gradually from 1965 to 1991 with the percentage of small-sized buildings dropped to 70% of the total number of private buildings.

Figure 4.4 and Table 4.5 further outline the growth in private buildings (non-domestic buildings and domestic buildings over 3 storeys in height) and the change of private buildings (in terms of number of building units) over the period of time from the Second World War till 2014. It is important to note the high speed of growth in terms of number of private buildings, particularly during the period from 1961 to 1966, with an average increase of more than 630 private buildings annually. Though with a lesser magnitude, the booming of private building production continued from 1967 to 1996 with an average increase of 400 to 560 per year. Contrastingly, following the world financial crisis in 1998, the output of private buildings reduced remarkably from 1997 to 2014.

Another essential feature which can be seen from Figure 4.5 is, in terms of number of building units, the size of private buildings had started to increase from 1977. Table 4.6 outlines the distribution (in terms of number of units) of private buildings. It is worth noting that out of the 397 (1.7% of the total stock) "extra-large-sized" buildings (i.e., buildings more than 401 units), there were 39 "jumbo-sized" buildings (i.e., buildings with more than 801 units), an equivalent to 0.2% of the total stock. Thus, the total number of building with units of 401 to 800 was 397 (1.7%) of the total stock.

Out of the total stock of 23,797 private buildings, the majority (15,350 or 64.5%) of private buildings in Hong Kong remain small in size (i.e., building with less than 50

Figure 4.4
Cumulative No. of Private Buildings (per Unit Group) under MBIS

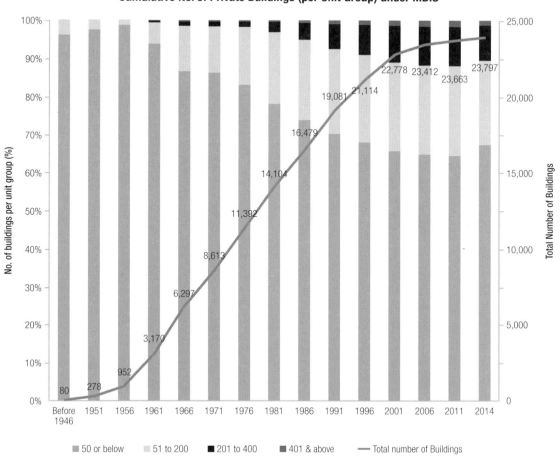

Note: For Non-Domestic Private Buildings and Domestic Buildings over Three Storeys
Source: Rating and Valuation Department, 2014

units). The second largest group was buildings with 51 to 200 units (5,626 or 23.6%). These two groups already amounted to over 88% of the total stock of private buildings in Hong Kong.

Figure 4.5 shows the cumulative number of private buildings and the five-year average annual growth rates over the past 50 years from pre-1946 to 2014. There were only 80 private buildings up to 1946. Due to the low base in terms of number of buildings, growth of private buildings drastically increased from the pre-1946 to 1966. The average annual growth rate over a five-year period for the years 1951, 1956, 1961 and 1966 were as high as 49.5%, 48.5%, 45.3% and 20.5% respectively. Along with the booming of population in Hong Kong from 1961 (3.13 Million) to 1996 (6.41 Million) (Census and Statistics

Table 4.5
Number of Private Buildings (Unit Distribution per Year Group) in Hong Kong
(As of 31 December 2014)

		No. of Private Buildings (Per Unit Groups)					
		50 or below	51 or 200	201 to 400	401 and above		
	Year	Small	Medium	Large	Extra Large	Total	Percentage (%)
1	2012-2014	68	48	18	0	134	0.6%
2	2007-2011	88	101	44	18	251	1.1%
3	2002–2006	210	175	195	54	634	2.7%
4	1997–2001	617	446	527	74	1,664	7.0%
5	1992–1996	953	612	411	57	2,033	8.5%
6	1987–1991	1,228	789	506	79	2,602	10.9%
7	1982–1986	1.158	821	342	54	2.375	10.0%
8	1977–1981	1,552	920	225	15	2,712	11.4%
9	1972–1976	2,041	678	51	9	2,779	11.7%
10	1967–1971	1,974	295	34	13	2,316	9.7%
11	1962–1966	2,545	569	56	20	3,190	13.4%
12	1957–1961	1,976	160	15	4	2,155	9.1%
13	1952–1956	669	5	0	0	674	2.8%
14	1947–1951	194	4	0	0	198	0.8%
15	≦1946	77	3	0	0	80	0.3%
	Total No. of Building (Per Unit Group)	15,350	5,626	2,424	397	23,797	
	Overall % in HK (Per Unit Group)	64.5%	23.6%	10.2%	1.7%		100.0%

Note : For Private Buildings Non-Domestic and Domestic Buildings over 3 storeys
Source: Home Affairs Department, 2014; Rating and Valuation Department, 2014

Table 4.6
Distribution of Private Buildings per Unit Groups
(As of 31 December 2014)

Group	Type	Building Size	No.	Percentage
1	Small	Building with total no. of unit ≤ 50	15,350	64.5%
2	Medium	Building with total no. of unit = 50 to 200	5,626	23.6%
3	Large	Building with total no. of unit = 201 to 400	2,424	10.2%
4	Extra Large	Building with total no. of unit ≥ 401	397	1.7%
		Total:	23,797	100.0%

Note : For Private Buildings Non-Domestic and Domestic Buildings over 3 storeys

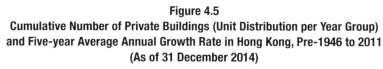

Figure 4.5
Cumulative Number of Private Buildings (Unit Distribution per Year Group)
and Five-year Average Annual Growth Rate in Hong Kong, Pre-1946 to 2011
(As of 31 December 2014)

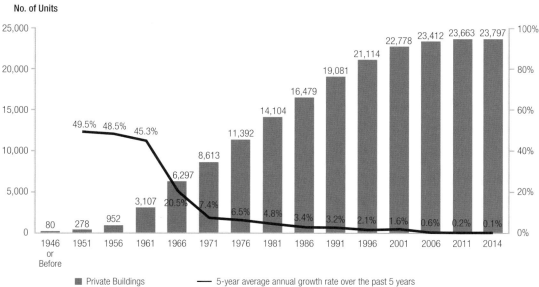

Department, 2012), i.e., about 2.1 times, the number of private buildings increased from 3,107 to 21,114. These figures show that the numbers of private buildings completed in 1996 was about 6.8 times more than that of 1961. The ratio in terms of the increase in numbers of private buildings between 1961 and 1996 is more than 3.2 times (6.8 times/2.1 times=3.2) when compared to the increase in population of Hong Kong. The reasons behind this higher increment ratio in private buildings than population are largely due to:

(a) Government policy in maintaining a stable environment for the sustainable and healthy development of the private property market; and

(b) The trend towards small household evidenced by the dropping of domestic household size from 4.4 in 1961 to 2.9 in 2011.

The five-year average annual growth rate of private buildings, however, started to plummet from 2001 to 2014. The five-year average annual growth rates in 2001, 2006 and 2011 and 2014 were only that of 1.6%, 0.6%, 0.2% and 0.1% respectively. Alternatively, the average annual number of new private buildings completed for the periods from 2002 to 2006, 2007 to 2011 and 2012 to 2014 were 634, 251 and 138 numbers respectively.

Figure 4.6
Growth of Private Building in Hong Kong from Pre-1946 to 2014
(in Building Unit Groups)

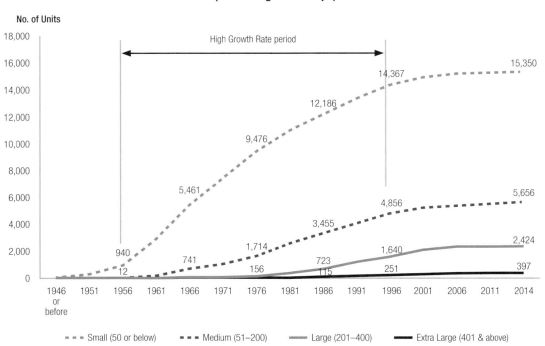

Figure 4.6 and Table 4.6 further illustrates the growth of different groups of private buildings from pre-1946 to 2014, with the rate of increase in small-sized private buildings much higher than the three other groups of private buildings. Mindful observation reveals that the high growth rate period of private buildings was from 1956 (952 in total) to 1996 (21,114 in total), in which over 500 private buildings were built annually.

Figures 4.6a to 4.6d outline the cumulative numbers of private buildings and a five-year average annual growth rates. Mindful observation reveals that the growth rate of small-sized private buildings (buildings with less than 50 units) is similar to that of private buildings of different sizes in Hong Kong. It is worth noting that, in 1946, there was a total of 77 small-sized buildings only. The number rocketed to 7,435 in 1971. Whilst the number of small-sized building grew to 15,350 in 2014, the 5-year average annual growth rates dropped continuously from 7.2% in 1971 to 0.2% in 2014.

There were only 3 medium-sized private buildings (buildings with 51 to 200 units) in 1946. In 15 years' time (i.e., in 1961), the number of this type of buildings increased to 172. But in 1966, the number of medium-sized private buildings rose sharply to 741. In 1996, the number continued to rise to 4,856. It is believed that the high growth rates in medium-sized buildings in these periods were mainly due to the combined effects of improved

Figure 4.6a
Cumulative Numbers of Small-sized Private Buildings (Building with units ≤ 50) and Average Annual Growth Rate in Hong Kong, Pre-1946 to 2014 (As at 31 December 2014)

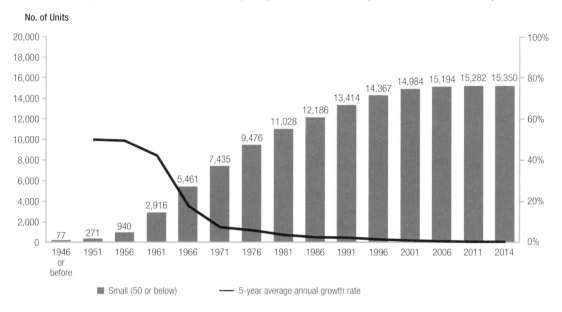

Figure 4.6b
Cumulative Numbers of Medium-sized Private Buildings (Building with units = 51 to 200) and Average Annual Growth Rate in Hong Kong, Pre-1964 to 2014 (As at 31 December 2014)

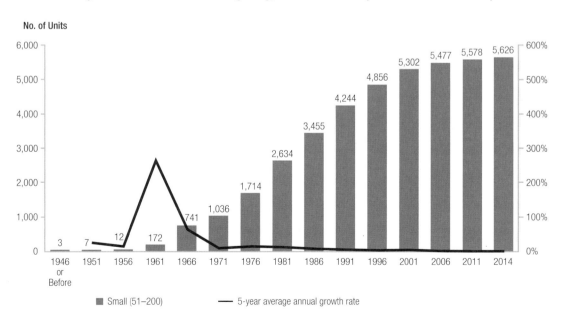

Figure 4.6c
Cumulative Numbers of Large-sized Private Buildings (Building with units = 201 to 400) and Average Annual Growth Rate in Hong Kong, Pre-1946 to 2014 (As at 31 December 2014)

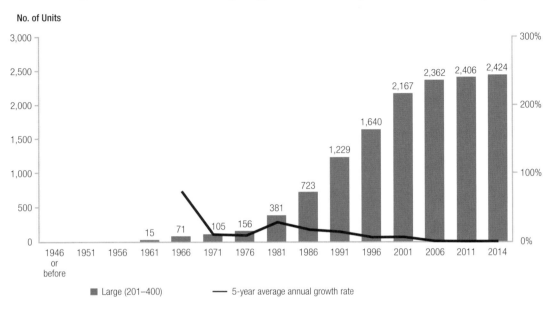

No. of Units

Large (201–400) ── 5-year average annual growth rate

Figure 4.6d
Cumulative Numbers of Extra-Large-sized Private Buildings (Building with units ≥ 401) and Average Annual Growth Rate in Hong Kong, Pre-1946 to 2014 (As at 31 December 2014)

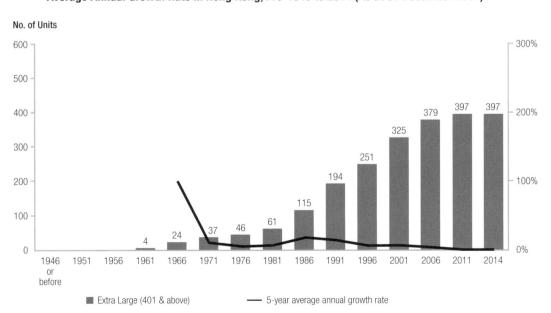

No. of Units

Extra Large (401 & above) ── 5-year average annual growth rate

technology in construction and the high land cost policy set by the Government. However, mindful observation reveals that, while the five-year average annual growth rate of medium-sized private buildings reduced from 1.8% in 2001 to 0.3% in 2014, it was already 1.5 times higher than that of the small-sized buildings with only 0.2% growth rate in 2014.

Records reveal that large-sized private buildings (buildings with 201 to 400 units), did not exist in 1956. But in 1961, there were 15 large-sized buildings, and the number continued to increase to 156 in 1976. In 1981, the number was sharply increased to 381. In 2001, the number continued to rise to 2,167. The five-year average annual growth rates in 1981, 1986, 1991, 1996 and 2001 were 28.8%, 18.0%, 14%, 6.7% and 6.4% respectively. Again, similar to the medium-sized buildings, it is believed that the high growth rates in large-sized buildings are mainly due to the combined effects of improved construction technology and the high land cost policy set by the Government.

For extra-large-sized private buildings (buildings with more than 401 units), similar to large-sized building, records reveal that no extra-large-sized building existed on or before 1956. There were only four extra-large-sized buildings completed during the period between 1957 and 1961. Whilst the number of extra-large-sized private buildings continued to grow in the next 20 years, there were only a total number of 61 in 1981. The five-year average annual growth rate between 1961 and 1981 was relatively low when compared with the three other groups of private buildings. However, mindful observation reveals that the five-year average annual growth rates for extra-large-sized buildings in 2006 were 3.3%. Such growth rates were already the highest among the total three groups of private buildings at that period. It is believed that the increase in number of extra-large-sized private buildings was due to the change in government policy by increasing the land sale size, enabling the construction of extra-large-sized private buildings.

From the analysis shown in Table 4.7, it is important to note that, from 2014 to 2024, not only will the number of private buildings falling under the MBIS increase (from 15,581 to 20,294), but also the number of different groups of private buildings will increase as well. In ten years' time, the number of small-sized buildings (Group 1) will increase by 19.1% to 14,094; medium-sized buildings (Group 2) by 44.8% to 1,389; large-sized buildings (Group 3) by 155% to 873 and extra-large-sized buildings (Group 4) by 211% to 188.

The above analysis further reveals that the number of buildings with medium- and large-sized flats is increased far more than buildings with small-sized flats. Assuming the policy in selection of the number of buildings by the Government remains unchanged at 2,000 per year, the time required by Registered Inspectors to conduct inspection and rectification works under the MBIS will inevitably increase as a result of the increase in size of private buildings (in terms of building units) along the time.

Table 4.7
Movement Analysis of Private Buildings falling under MBIS from 2014 to 2024

	No. of Private Buildings (Per Unit Groups)					
	50 or below	51 to 200	201 to 400	401 and above		
	Small	Medium	Large	Extra Large		
Year	Group 1	Group 2	Group 3	Group 4	Total	Percentage (%)
Total No. of Private Buildings under MBIS Year 2014 [#1]	11,831	3,098	563	89	15,581	Private Buildings Buildings completed on or before 1984
	75.9%	19.9%	3.6%	0.6%		
Total No. of Private Buildings under MBIS by Year 2024	14,094	4,487	1,436	277	20,294	Private Buildings Buildings completed on or before 1994
	69.4%	22.1%	7.1%	1.4%		
Growth on the no. of private buildings falling under the MBIS (2014 to 2024)	2,263 (19.1%)	1,389 (44.8%)	873 (155%)	188 (211%)	4,713	Movement

Note: For Private Buildings Non-Domestic and Domestic Buildings over 3 storeys
[#1]: The number of buildings are captured at 31 December 2014

4.3 Summary of Private Buildings under the Mandatory Building Inspection Scheme

As of 31 December 2014, out of the total 43,163 private buildings across the Hong Kong territory, there were 23,797 private non-domestic buildings and domestic buildings over 3 storeys in height, 64.5% of the total stock was small in size (i.e., buildings with less than 50 units). The second largest group, 23.6% of the total stock, was buildings with 51 to 200 units. These two groups already account for over 88% of the total stock of private buildings in Hong Kong. Among them, 15,581 buildings have already reached the age of 30 years or above. These aged buildings are governed by the MBIS, implemented on 30 June 2012 by the Buildings Department. This figure represents almost 66% of the total stock of private buildings. It is forecasted that, in 10 years' time (i.e., 2024), the number of aged buildings under the MBIS will drastically increase to 20,294, equivalent to over 85% of the total stock of private buildings in Hong Kong.

Among the three regions within the Hong Kong territory, Kowloon was the most densely populated in terms of number of private buildings falling under the MBIS. Hong Kong Island ranked the second, with 5,829 buildings or 73.2% of its total stock of buildings in this region. New Territories and Islands was the least dense region among the three, with 2,466 buildings or 37.6% of its total stock. However, in 10 years' time (i.e., in 2024), whilst the rankings of the densest private buildings falling under the MBIS will remain unchanged, but in the rate of ageing in private buildings will be dramatically increased in the New Territories and Islands from 37.6% in 2014 to 71.8% in 2024.

Among the 18 districts in Hong Kong, Central and Western had the largest number of private buildings—as many as 3,162 in numbers. Yau Tsim Mong ranked the second, with 3,041 buildings. Sai Kung and the Islands ranked the penultimate and the last, with 480 and 303 in number respectively.

Yau Tsim Mong was the "densest aged" district falling under the MBIS, with 2,478 aged private buildings (81.5% of its total stock of private buildings) as of 31 December 2014. However, in 2024, although the number of private buildings falling under the MBIS in Yau Tsim Mong will increase to 2,738 (90% of its stock of private buildings), its first position in terms of number of aged private buildings falling under the MBIS will be replaced by Central and Western District, with 2,902 buildings. Kowloon City will replace Yau Tsim Mong as the densest aged district; with 91.8% of its stock of private buildings will reach the age of 30 years or above.

Sai Kung was the "youngest" district in terms of density of aged buildings falling under the MBIS. Out of its total number of 480 buildings, only 55 reached the age of 30 or above, i.e., 11.5%.

Finally, the rates of ageing on the Islands, Shatin and Tuen Mun in the coming years will become remarkably high. The density on the Islands, Shatin, and Tuen Mun will increase from 30.4%, 25.9% and 32.4% to 79.9%, 67% and 83.8% in year 2024 respectively.

4.4 Conclusions

As of 31 December 2014, there were 43,163 private buildings across the Hong Kong, among which 23,797 were non-domestic buildings and domestic buildings over 3 storeys in height. These buildings will be governed by the MBIS and MWIS, once their ages reach 30 years or above. There were a total of 15,581 private buildings (65.5%) that reached the age of 30 years or above, and they will be governed by the MBIS or MWIS. It is foreseeable that the problem of aged private buildings will become more severe in the coming years when there will be as many as 20,294 (85.3%) private buildings falling under the MBIS and MWIS in 2024.

Analyses also reveal that the majority of private buildings in Hong Kong are small in size (i.e., buildings with less than 50 units). Out of the total stock of 23,797 private buildings, 15,350 buildings or 64.5% of the total stock belong to this group. 5,626 buildings or 23.6% of the total stock, the second largest group of private buildings, are buildings with 51 to 200 units. These first two groups already amount to 88.1% of the total stock of private buildings in Hong Kong. Analyses further reveal that the distribution of private buildings (in terms of number of building units) under the MBIS is shifting from small-sized buildings to large-sized buildings. Assuming the policy in the selection of the number of buildings by the Government remains unchanged at 2,000 per year, the time required by Registered Inspectors to conduct the inspection and rectification works under the MBIS will inevitably increase as a result of the rise in size of private buildings (in terms of number of building units) along the time.

References

Buildings Department (2012a). *Mandatory Building Inspection Scheme (Pamphlet on MBIS)*. Hong Kong: Buildings Department of HKSAR.

Census and Statistics Department (2012). *Hong Kong in Figures*. Hong Kong: Census and Statistics Department of HKSAR.

Home Affairs Department (2014). *Database of Private Buildings in Hong Kong.* Home Affair of HKSAR. Available at https://bmis.buildingmgt.gov.hk/eng/index.php (Accessed on 31 December 2014).

Rated and Valuation Department (2014). *Hong Kong Property Review*. Hong Kong: Rated and Valuation Department of HKSAR.

Tsui, Ho (2012)，九龍城屋宇維修研討會2012，建設健康九龍城協會，英國特許建造學會（香港) 及九龍城民政事務處，香港。

CHAPTER 5

Analyses of
Private Buildings
under MBIS
in the 18 Districts

5.1 Introduction

Private buildings in the 18 District Council districts of Hong Kong under the MBIS are critically examined in respect of their key features and sizes, as well as expression in number of building units and percentage (%) per year group. Further analyses of the change in numbers, compositions of different sizes and the rate of ageing of private buildings under the MBIS from 2014 to 2024 are carried out.

5.2 Private Buildings under MBIS in District Council Districts

In order to capture the changes in number of private buildings that fell or will fall under the MBIS across the 18 District Council districts from 2014 to 2024, this section provides a critical examination on the age of private buildings in each of the 18 districts from Pre-war years to 2024 as follows:

(a) Cumulative number of private buildings;

(b) Percentage of unit group distributions in groups of five-year interval;

(c) Number of private buildings in unit group distributions in groups of five-year interval;

(d) Total number of private buildings and their ranking in 2014 as compared with the total number of private buildings in Hong Kong;

(e) Total number of private buildings falling under the region of the MBIS and their ranking in 2014 and 2024 respectively;

(f) Percentage of private buildings falling under the MBIS in 2014 and 2024, as compared with the total number of private buildings in the corresponding district; and

(g) Ageing rate and percentage of private building falling under the MBIS from 2014 to 2024.

Central and Western District, Hollywood Road

5.2.1   Central and Western District

Key Features

(a)   Ranked first in 2014 in number of private buildings with a total of 3,162 private buildings in the district

(b)   Ranked second in 2014 with 2,412 (76.3%) private buildings falling under the MBIS

(c)   Shall rank first in 2024 with 2,902 (91.8%) private buildings falling under the MBIS

(d)   Changes in 10 years' time

  • Ranking shall move one notch in terms of number of private buildings falling under the MBIS

  • Projected number of private buildings under the MBIS shall increase by 490

  • Shall rank no. 13 in terms of ageing rate (15.5%) of private buildings under the MBIS

Figure 5.1
Cumulative No. of Buildings (Per Unit Group) in Central and Western District
(As of 31 December 2014)

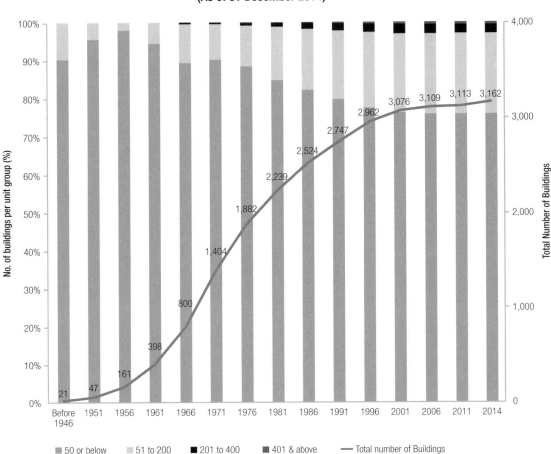

Note: For Non-Domestic Private Buildings and Domestic Buildings over 3 Storeys

Comment

According to the Central and Western District Council (District Council, 2014), "the Central and Western District, one of the earliest developed areas in the territory, is a commercial, financial, legal and political centre of Hong Kong. Up to mid-2006, the district has a residential population of about 259,300. Over 90% live in private buildings, varied from luxury homes at the Peak and on Mid-levels to old tenements in Sheung Wan and Sai Wan.

Central and Western District, with 3,162 private buildings ranked first among the 18 districts in 2014 in terms of number of private buildings. As of 31 December 2011, there was a total of 2,412 (76.3%) private buildings, falling under the MBIS in the district, and

Table 5.1
Number of Private Buildings (Unit Distribution Per Year Group)
in Central and Western District(As of 31 December 2014)

	Year	No. of Private Buildings (Per Unit Groups)					Percentage (%)
		50 or above Small	51 or 200 Medium	201 to 400 Large	401 and above Extra Large	Total	
1	2012–2014	43	6	0	0	49	1.5%
2	2007–2011	2	2	0	0	4	0.1%
3	2002–2006	13	12	5	3	33	1.0%
4	1997–2001	51	49	8	6	114	3.6%
5	1992–1996	107	89	19	0	215	6.8%
6	1987–1991	113	97	13	0	223	7.1%
7	1982–1986	176	90	15	4	285	9.0%
8	1977–1981	233	114	10	0	357	11.3%
9	1972–1976	396	73	7	2	478	15.1%
10	1967–1971	554	47	3	0	604	19.1%
11	1962–1966	338	61	3	0	402	12.7%
12	1957–1961	218	19	0	0	237	7.5%
13	1952–1956	113	1	0	0	114	3.6%
14	1947–1951	26	0	0	0	26	0.8%
15	≦1946	19	2	0	0	21	0.7%
	Total No. of Buildings (Per Unit Group)	2,402	662	83	15	3,162	
	% in Central & Western (Per Unit Group)	76%	20.9%	2.6%	0.5%		100.0%
	Total No. of Buildings under MBIS Year 2014 and % in terms of no. of buildings in 2014	2,012	364	31	5	2,412	Ranked No. 2 (2014)
		63.6%	11.5%	1.0%	0.2%	76.3%	
	Total No. of Buildings under MBIS by Year 2024 and % in terms of no. of buildings in 2014	2,368	438	38	58	2,902	Ranked No. 1 (2024)
		74.9%	13.9%	1.2%	1.8%	91.8%	
	Private Buildings under MBIS (Movement 2014–2024)	356	74	7	53	490	Ranking 1 notch up
		11.3%	2.3%	0.2%	1.7%	15.5%	

Note: For Private Buildings Non-Domestic and Domestic Buildings over 3 storeys

was second among the 18 districts. In this aspect, the next 10 years, the projected number shall further increase to 2,902 (91.8%), and the Central and Western District shall rank first among all 18 districts.

In 1946, there were 21 private buildings in the district in 1946 which already represented 26% all private buildings in Hong Kong. In the period from 1957 to 1986, 2,363 private buildings (75.9%) were completed. In terms of size distribution (number of building units) of private buildings in the district, two characteristics are observed. First, the proportion of small-sized buildings was relatively high. As compared with the average of 64.5% in the territory, the proportion of small-sized private buildings in the district was 76.0%. Second, the proportion of large-sized private buildings was relatively lower. As compared with the average of 10.2% among all districts, the proportion of large-sized private buildings in the district was merely 2.6%.

As of 31 December 2014, out of the 2,412 (76.3%) private buildings falling under the MBIS, 2,012 (63.6%) were small-sized buildings, 364 (11.5%) medium-sized buildings, 31 (1%) large-sized buildings and 5 (0.2%) extra-large-sized buildings. In 2024, the number of aged private buildings falling under the MBIS shall increase to 2,902 (91.8%). This represents a net increase of 490 (15.5%) private buildings. The distributions of aged private buildings falling under the MBIS shall be 2,368 (74.9%) small-sized buildings, 438 (13.9%), medium-sized buildings, 38 (1.2%), large-sized buildings and 58 (1.8%) extra-large-sized buildings. The ranking of the district, in terms of ageing rate (15.5%) of aged private buildings from 2014 to 2024, shall be no. 13.

Yam Tsim Mong District, Shanghai Street

5.2.2 Yau Tsim Mong District

Key Features

(a) Ranked second in 2014 in number of private buildings with a total of 3,041 private buildings

(b) Ranked first in 2014 with 2,478 private buildings (81.5%) falling under the MBIS

(c) Shall rank second in 2024 with 2,638 (90%) private buildings falling under the MBIS

(d) Movements in 10 years' time

- Ranking shall move one notch down in terms of number of private buildings falling under the MBIS

- Projected number of buildings under the MBIS shall increase by 260

- Shall rank the lowest in terms of ageing rate (8.5%) of buildings under the MBIS

Figure 5.2
Cumulative No. of Buildings (Per Unit Group)
in Yau Tsim Mong District (As of 31 December 2014)

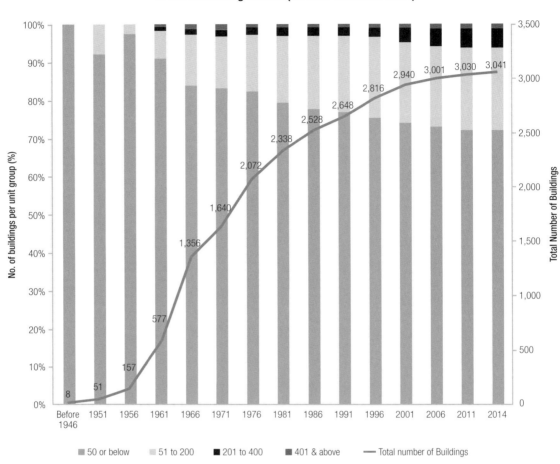

Note: For Non-Domestic Private Buildings and Domestic Buildings over 3 Storeys

Comment

According to the Yau Tsim Mong District Council (District Council, 2014), "it is the focal point of traffic arteries in Hong Kong. The major transport facilities in the district include the MTR Hung Hom Station (the terminus of East Rail Line and West Rail Line), Hung Hom Cross Harbour Tunnel, West Harbour Crossing, the Star Ferry, Austin Road Cross Boundary Coach Terminus and the Hong Kong China Ferry Terminal in Tsim Sha Tsui. Economic activities in Yau Tsim Mong District are mainly supported by commercial activities and supplemented by tourism and light industries".

Yau Tsim Mong District, with 3,041 private buildings, ranked second in 2014 in terms of number of private buildings among the 18 districts. As of 31 December 2014, there was

Table 5.2
Number of Private Buildings (Unit Distribution Per Year Group)
in Yau Tsim Mong District (As of 31 December 2014)

| Year of Completion | | No. of Private Buildings (Per Unit Groups) | | | | | Percentage (%) |
| | | 50 or below | 51 to 200 | 201 to 400 | 401 and above | Total | |
		Small	Medium	Large	Extra Large		
1	2012–2014	2	7	2	0	11	0.4%
2	2007–2011	2	15	12	0	29	1.0%
3	2002–2006	11	15	31	4	61	2.0%
4	1997–2001	54	25	41	4	124	4.1%
5	1992–1996	87	70	9	2	168	5.5%
6	1987–1991	73	44	3	0	120	3.9%
7	1982–1986	106	76	6	2	190	6.2%
8	1977–1981	153	100	12	1	266	8.7%
9	1972–1976	343	81	8	0	432	14.2%
10	1967–1971	225	47	7	5	284	9.3%
11	1962–1966	613	137	16	13	779	25.6%
12	1957–1961	372	39	5	4	420	13.8%
13	1952–1956	106	0	0	0	106	3.5%
14	1947–1951	39	4	0	0	43	1.4%
15	≦1946	8	0	0	0	8	0.3%
Total No. of Buildings (Per Unit Group)		2,194	660	152	35	3,041	
% in Yau Tsim Mong (Per Unit Groups		72.1%	21.7%	5.0%	1.2%		100.0%
Total No. of Buildings under MBIS Year 2014 and % in terms of no. of buildings in 2014		1,948	454	52	24	2,478	Ranked No. 1 (2014)
		64.1%	14.9%	1.7%	0.8%	81.5%	
Total No. of Buildings under MBIS by Year 2024 and % in terms of no. of buildings in 2014		2,086	564	62	26	2,738	Ranked No. 2 (2024)
		68.6%	18.6%	2.0%	0.9%	90.0%	
Private Buildings under MBIS (Movement 2014–2024)		138	110	10	2	260	Ranking 18 notch down
		4.5%	4.0%	0.3%	0.1%	8.5%	

Note: For Private Buildings Non- Domestic and Domestic Buildings over 3 storeys

a total of 2,478 (81.5%) private buildings in the district falling under the MBIS, ranked first among the 18 districts in Hong Kong. In this aspect, in the next 10 years, i.e. by 2024, the projected number shall further increase to 2,738 (90%), and the district shall be move one down notch to rank second among all districts.

In 1946, there was a total of eight private buildings in the district, which represented only 10% of the total number of private buildings in Hong Kong. In the period from 1957 to 1976, 1,915 private buildings (63.2%) were completed. It is worth special notice that 779 aged private buildings (25.7%) in the district were completed during the period 1962 to 1966, with 13 of them being extra-large-sized private buildings. In terms of size distribution (number of building units) of private buildings in the district, two characteristics are observed. First, the proportion of small-sized buildings was relatively high. As compared with the average of 64.5% in the territory, the proportion of small-sized private buildings in the district was 72.1%. Second and conversely, the proportion of large-sized private buildings was relatively lower. Compared with the average of 10.2% in the territory, the proportion of large-sized private buildings in the district was only 5.0%.

As of 31 December 2014, out of the total 2,478 (81.5%) private buildings under the MBIS in the district, 1,948 (64.1%) were of small-sized buildings, 454 (14.9%) medium-sized buildings, 52 (1.7%) large-sized buildings and 24 (0.8%) extra-large-sized buildings. In the next 10 years, the number of aged private buildings under the MBIS in the district shall increase to 2,738 (90%). This represents a net increase of 260 (8.3%) private buildings. The distribution of aged private buildings under the MBIS in the district shall be 2,086 (68.6%) small-sized buildings, 564 (18.6%) medium-sized buildings, 62 (2%) large-sized buildings and 26 (0.9%) extra-large-sized buildings. The ranking, in terms of ageing rate (8.5%) of aged private buildings from 2014 to 2024 shall be the lowest.

Kowloon City District, Hau Wong Road

5.2.3   Kowloon City District

Key Features

(a)   Ranked third in 2014 in number of private buildings with a total of 2,629 private buildings in the district

(b)   Ranked third in 2014 with 2,017 (76.7%) private buildings falling under the MBIS

(c)   Shall rank third in 2024 with 2,345 (89.2%) private buildings falling under the MBIS

(d)   Changes in 10 years' time

- Ranking shall remain unchanged in terms of number of private buildings falling under the MBIS

- Projected number of buildings under the MBIS shall increase by 277

- Shall rank 17th in terms of ageing rate (10.5%) of private buildings under the MBIS.

Figure 5.3
Cumulative No. of Buildings (Per Unit Group)
in Kowloon City District (As of 31 December 2014)

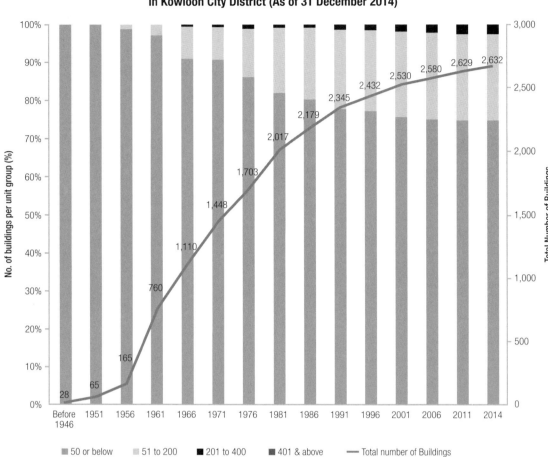

Note: For Non-Domestic Private Buildings and Domestic Buildings over 3 Storeys

Comment

The Kowloon City District Council (District Council, 2014) highlights that "covering an area of approximately 1,000 hectares, the Kowloon City District is chiefly a residential area accommodating a population of about 374,300. The majority of its population dwells in private sector housing, including old tenement buildings, private residential developments and low-rise villas. The Kowloon City District is a very distinctive place. Its Kowloon Walled City was one of the earliest developments in the territory. In 1993, the Walled City was reconstructed as the Walled City Park where many relics of historical value have been preserved".

Table 5.3
Number of Private Buildings (Unit Distribution Per Year Group)
in Kowloon City District (As of 31 December 2014)

	Year of Completion	No. of Private Buildings (Per Unit Groups)					Percentage (%)
		50 or above	51 or 200	201 to 400	401 and above	Total	
		Small	Medium	Large	Extra Large		
1	2012–2014	0	3	0	0	3	0.1%
2	2007–2011	29	8	12	0	49	1.9%
3	2002–2006	24	19	7	0	50	1.9%
4	1997–2001	33	53	9	3	98	3.7%
5	1992–1996	55	27	5	0	87	3.3%
6	1987–1991	76	77	13	0	166	6.3%
7	1982–1986	92	69	1	0	162	6.2%
8	1977–1981	189	122	0	0	314	11.9%
9	1972–1976	153	75	7	0	255	9.7%
10	1967–1971	303	29	6	0	338	12.8%
11	1962–1966	272	73	4	1	350	13.3%
12	1957–1961	574	20	1	0	595	22.6%
13	1952–1956	98	2	0	0	100	3.8%
14	1947–1951	37	0	0	0	37	1.4%
15	≦1946	28	0	0	0	28	1.1%
Total No. of Buildings (Per Unit Group)		1,963	597	65	4	2,629	
% in Kowloon City (Per Unit Group)		74.6%	22.7%	2.5%	0.2%		100.0%
Total No. of Buildings under MBIS Year 2014 and % in terms of no. of buildings in 2014		1,727	384	19	1	2,131	Ranked No. 3 (2014)
		65.6%	14.6%	0.7%	0.0%	81.1%	
Total No. of Buildings under MBIS by Year 2024 and % in terms of no. of buildings in 2014		1,864	508	35	1	2,408	Ranking No. 3 (2024)
		70.8%	19.3%	1.3%	0.0%	91.6%	
Private Buildings under MBIS (Movement 2014–2024)		137	124	16	0	277	Ranking remains unchanged
		5.2%	4.7%	0.6%	0.0%	10.5%	

Note: For Private Buildings Non-Domestic and Domestic Buildings over 3 storeys

The Kowloon City District, with 2,629 private buildings, ranked third in 2014 in terms of number of private buildings among the 18 districts. As of 31 December 2014, there was a total of 2,131 (81.1%) private buildings in the district falling under the MBIS, ranking third among the 18 districts. In this aspect, in the next 10 years, i.e. by 2024, the projected number shall further increase to 2,408 (91.6%) and its ranking shall remain unchanged among the districts.

In 1946, there were only 28 private buildings in the district, representing 35% of the total number of private buildings in Hong Kong. In the period from 1957 to 1981, 1,852 private buildings (70.4%) were completed. Another essential feature of Kowloon City is that 97.3% of the private buildings in the district are of small-sized to medium-sized private buildings and that the buildings were relatively low in level. This was largely due to the fact that the Kowloon City district was situated in the landing pathway of the Hong Kong Kai Tak Airport.

As of 31 December 2014, out of the 2,131 (81.1%) private buildings in the district under the MBIS, 1,727 (65.6%) were of small-sized buildings, 384 (14.6%) medium-sized buildings, 19 (0.7%) large-sized buildings and 1 (0.0%) extra-large-sized buildings. In 2024, the number of aged private buildings in the district under the MBIS shall increase to 2,408 (91.6%). This represents a net increase of 277 (10.5%) private buildings. The distribution of aged private buildings falling under the MBIS will be 1,864 (70.8%) small-sized buildings, 508 (19.3%) medium-sized buildings, 35 (1.3%) large-sized buildings and 1 (0.0%) extra-large-sized buildings. The ranking, in terms of ageing rate (0.5%) of aged private buildings from 2011 to 2021, shall be the 17th.

Sham Shui Po District, Fuk Wing Street

5.2.4   Sham Shui Po District

Key Features

(a)   Ranked fourth in 2014 in number of private buildings with a total of 2,081 private buildings in the district

(b)   Ranked fourth in year with 1,660 (79.8%) buildings falling under the MBIS

(c)   Shall rank fifth in 2024 with 1,888 (90.7%) buildings falling under the MBIS

(d)   Changes in 10 years' time

- Ranking shall move one notch down in terms of number of private buildings falling under the MBIS

- Projected number of buildings under the MBIS shall increase by 228

- Shall rank 16th in terms of ageing rate (11%) of private buildings under the MBIS

Figure 5.4
Cumulative No. of Buildings (Per Unit Group)
in Sham Shui Po District (As of 31 December 2014)

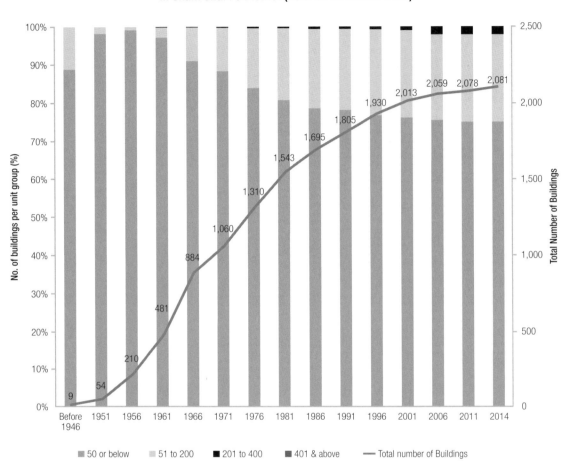

Note: For Non-Domestic Private Buildings and Domestic Buildings over 3 Storeys

Comment

The Sham Shui Po District Council (District Council, 2014) highlights that "Sham Shui Po District was already densely populated back in the 1950s and 1960s. Its population once reached 440,000 in 1986 and is now 365 600. There is a considerable number of private blocks, built in the 1950s and 1960s, along Cheung Sha Wan Road, Lai Chi Kok Road and Tai Po Road. Most of them are five to seven-storey buildings. Some other private housing developments in the district, like Mei Foo Sun Chuen and Yau Yat Tsuen, have marked features. The Sham Shui Po District is probably Hong Kong's earliest industrial and commercial centre. Wholesale and retail businesses of mainly textile and

Table 5.4
Number of Private Buildings (Unit Distribution Per Year Group)
in Sham Shui Po District (As of 31 December 2014)

Year of Completion		No. of Private Buildings (Per Unit Groups)					Percentage (%)
		50 or above	51 or 200	201 to 400	401 and above		
		Small	Medium	Large	Extra Large	Total	
1	2012–2014	0	1	2	0	3	0.1%
2	2007–2011	5	10	4	0	19	0.9%
3	2002–2006	22	3	19	2	46	2.2%
4	1997–2001	55	23	5	0	83	4.0%
5	1992–1996	68	55	2	0	125	6.0%
6	1987–1991	77	31	2	0	110	5.3%
7	1982–1986	87	63	2	0	152	7.3%
8	1977–1981	147	84	2	0	233	11.2%
9	1972–1976	163	84	3	0	250	12.0%
10	1967–1971	131	45	0	0	176	8.5%
11	1962–1966	338	64	1	0	403	19.4%
12	1957–1961	260	10	1	0	271	13.0%
13	1952–1956	155	1	0	0	156	7.5%
14	1947–1951	45	0	0	0	45	2.2%
15	≦1946	8	1	0	0	9	0.4%
Total No. of Buildings (Per Unit Group)		1,561	475	43	2	2,081	
% in Sham Shui Po (Per Unit Groups)		75.0%	22.8%	2.1%	0.1%		100.0%
Total No. of Buildings under MBIS Year 2014 and % in terms of no. of buildings in 2014		1,323	329	8	0	1,660	Ranked No. 4 (2014)
		63.6%	15.8%	0.4%	0.0%	79.8%	
Total No. of Buildings under MBIS by Year 2024 and % in terms of no. of buildings in 2014		1,461	415	12	0	1,888	Ranked No. 5 (2024)
		70.2%	19.9%	0.6%	0.0%	90.7%	
Private Buildings under MBIS (Movement 2014–2024)		138	86	4	0	228	Ranking 1 notch down
		6.6%	4.2%	0.2%	0.0%	11%	

Note: For Private Buildings Non-Domestic and Domestic Buildings over 3 storeys

clothing, apparels, piece goods and non-staple food, are concentrated in Cheung Sha Wan and Lai Chi Kok".

The Sham Shui Po District, with 2,081 private buildings ranked fourth in 2014 in terms of number of private buildings among the 18 districts. As of 31 December 2014, there was a total of 1,660 (79.8%) private buildings in the district falling under the MBIS, making the district rank fourth among the 18 districts. In this aspect, in the next 10 years, the projected number shall increase to 1,888 (90.7%) and shall rank fifth among the districts.

In 1946, there were only nine private buildings in the district which represented 11.2% of the total number of private buildings in Hong Kong. In the period from 1952 to 1986, 1,641 private buildings (79.0%) were completed. Between 1962 and 1966, 403 aged private buildings (19.4%) were completed and there are no extra-large-sized private buildings in the district. It was not until between 2002 and 2006 that two extra-large-sized private buildings were built.

As of 31 December 2014, out of the 1,660 (79.8%) numbers of private buildings falling under the MBIS, 1,323 (63.6%) were small-sized buildings, 329 (15.8%) medium-sized buildings but only 8 (0.4%) large-sized buildings. In 2024, the numbers of aged private buildings falling under the MBIS shall increase to 1,888 (90.7%). This represents a net increase of 228 private buildings (11%). The distribution of aged private buildings falling under the MBIS shall be 1,461 (70.2%) small-sized buildings, 415 (19.9%) medium-sized buildings and 12 (0.6%) large-sized buildings. The ranking, in terms of ageing rate (11%) of aged private buildings from 2014 to 2024, shall be the 16th.

Wai Chai District, Queen's Road East

5.2.5   Wan Chai District

Key Features

(a)   Ranked fifth in 2014 in number of private buildings with a total of 2,071 private buildings in the district

(b)   Ranked fifth in 2014 with 1,621 (78.3%) falling under the MBIS

(c)   Shall rank fourth in 2024 with 1,931 (93.2%) private buildings falling under the MBIS

(d)   Changes in 10 years' time

- Ranking shall move one notch down in terms of number of private buildings falling under the MBIS

- Projected number of buildings under the MBIS shall increase by 310

- Shall rank 14th in terms of ageing rate (15%) of private buildings under the MBIS

Figure 5.5
Cumulative No. of Buildings (Per Unit Group)
in Wan Chai District (As of 31 December 2014)

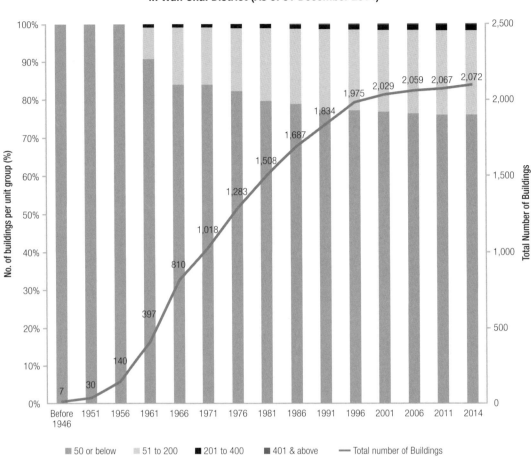

Note: For Non-Domestic Private Buildings and Domestic Buildings over 3 Storeys

Comment

The Wan Chai District Council (District Council, 2014) highlights that "Wan Chai was originally a fishing village. Throughout the years, it has developed from a residential area to a prime centre of business, conventions, exhibitions, cultural and sports activities, entertainment and shopping and is known for its unique and harmonious blend of old traditions and new developments. It is one of the oldest districts of Hong Kong with rich cultural heritage and traditions".

Wan Chai District, with 2,071 private buildings ranked fifth in 2014 in terms of number of private buildings among the 18 districts. As of 31 December 2014, there was a total of

Table 5.5
Number of Private Buildings (Unit Distribution per Year Group)
in Wan Chai District (As of 31 December 2014)

		No. of Private Buildings (Per Unit Groups)					
		50 or above	51 or 200	201 to 400	401 and above		Percentage (%)
	Year of Completion	Small	Medium	Large	Extra Large	Total	
1	2012–2014	1	3	0	0	5	0.2%
2	2007–2011	3	4	1	0	8	0.4%
3	2002–2006	12	16	2	0	30	1.4%
4	1997–2001	35	18	1	0	54	2.6%
5	1992–1996	103	33	4	1	141	6.8%
6	1987–1991	90	53	3	1	147	7.1%
7	1982–1986	127	49	2	1	179	8.6%
8	1977–1981	146	72	6	1	225	10.9%
9	1972–1976	199	61	4	1	265	12.8%
10	1967–1971	174	33	1	0	208	10.0%
11	1962–1966	320	89	4	0	413	19.9%
12	1957–1961	220	33	4	0	257	12.4%
13	1952–1956	110	0	0	0	110	5.3%
14	1947–1951	23	0	0	0	23	1.1%
15	≦1946	7	0	0	0	9	0.3%
Total No. of Building (Per Unit Group)		1,570	464	32	5	2,071	
% in Wan Chai (Per Unit Group)		75.8%	22.4%	1.5%	0.2%		100.0%
Total No. of Buildings under MBIS Year 2011 and % in terms of no. of building in 2014		1,281	317	20	3	1,621	Ranked No. 5 (2014)
		61.8%	15.3%	1.0%	0.1%	78.3%	
Total No. of Buildings under MBIS by Year 2024 and % in terms of no. of building in 2014		1,488	412	26	5	1,931	Ranked No. 4 (2024)
		71.8%	19.9%	1.3%	0.2%	93.2%	
Private Buildings under MBIS (Movement 2014–2024)		207	95	6	2	310	Ranking 1 notch up
		10.0%	4.6%	0.3%	0.1%	15%	

Note: For Private Buildings Non-Domestic and Domestic Buildings over 3 storeys

1,621 (78.3%) private buildings in the district falling under the MBIS, ranking fifth among the 18 districts. In this aspect, in the next 10 years, the projected number shall increase to 1,931 (93.2%), and the district shall be rank fourth among the 18 districts.

In 1946, there were only seven private buildings in the district representing 8.8% of the total number of private buildings in Hong Kong. In the period from 1957 to 1986, 1,547 private buildings were completed (74.8%). It is also worth noting that 413 aged private buildings (20.0%) were completed during the period from 1962 to 1966. Moreover, there were only five extra-large-sized private buildings in the district. As compared with the total number of 397 extra-large-sized private buildings in the territory, this number is relatively low.

As of 31 December 2014, out of the 1621 (78.3%) private buildings falling under the MBIS, 1,281 (61.8%) were small-sized buildings, 317 (15.3%) medium-sized buildings, 20 (1%) large-sized buildings and 3 (0.1%) extra-large-sized buildings. In this aspect, the projected numbers in 2024 will increase to 1,931 (93.2%). This represents a net increase of 310 private buildings (15%). The distribution of aged private buildings falling under the MBIS shall be 1,488 (71.8%) small-sized buildings, 412 (19.9%) medium-sized buildings, 26 (1.3%) large-sized buildings and 5 (0.2%) extra-large-sized buildings. The ranking, in terms of ageing rate (15%) of aged private buildings from 2014 to 2024, shall be the 14th.

Eastern District, King's Road

5.2.6   Eastern District

Key Features

(a) Ranked sixth in 2014 in number of private buildings with a total of 1,728 private buildings in the district

(b) Ranked sixth in 2014 with 1,176 (68.1%) buildings falling under the MBIS

(c) Shall rank sixth in 2024 with 1,559 (90.2%) private buildings falling under the MBIS

(d) Changes in 10 years' time

   • Ranking shall remain unchanged in terms of number of private buildings falling under the MBIS

   • Projected number of buildings under the MBIS shall increase by 383

   • Shall be ranked no. 11 in terms of ageing rate (22.2%) of private buildings under the MBIS

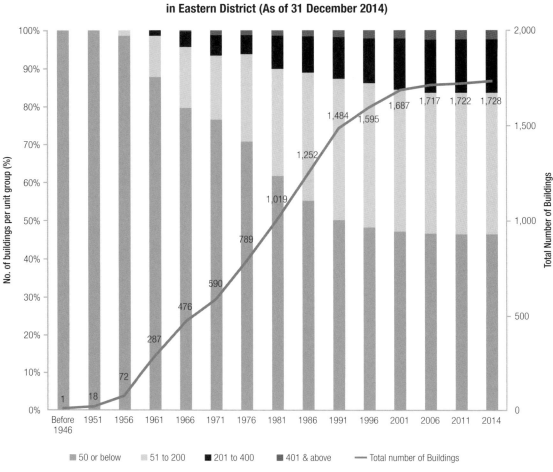

Figure 5.6
Cumulative No. of Buildings (Per Unit Group)
in Eastern District (As of 31 December 2014)

Note: For Non-Domestic Private Buildings and Domestic Buildings over 3 Storeys

Comment

The Eastern District was originally a backwater of fishing villages, quarries and dockyards and has turned into a mixture of residential, industrial areas, commercial and large shopping malls. The Eastern District Council (District Council, 2014) highlights that "The Eastern District is well-served by Mass Transit Railway (MTR), buses, trams, ferries and green mini-bus (GMB) services. The opening of the Eastern Harbour Crossing since 1989 has significantly improved traffic linkage across the harbour. There are 15 bus termini or public transport interchanges in the district, providing various bus and GMB routes for the Eastern District residents to and from Hong Kong, Kowloon and the New Territories".

Table 5.6
Number of Private Buildings (Unit Distribution Per Year Group)
in Eastern District (As of 31 December 2014)

	Year of Completion	No. of Private Buildings (Per Unit Groups)					Percentage (%)
		50 or above	51 or 200	201 to 400	401 and above	Total	
		Small	Medium	Large	Extra Large	Total	
1	2012–2014	1	4	1	0	6	0.3%
2	2007–2011	0	5	0	0	5	0.3%
3	2002–2006	2	10	12	6	30	1.7%
4	1997–2001	27	25	38	2	92	5.3%
5	1992–1996	28	49	26	8	111	6.4%
6	1987–1991	51	133	42	6	232	13.4%
7	1982–1986	62	135	30	6	233	13.5%
8	1977–1981	70	105	51	4	230	13.3%
9	1972–1976	107	82	6	4	199	11.5%
10	1967–1971	73	23	12	6	114	6.6%
11	1962–1966	127	45	16	1	189	10.9%
12	1957–1961	181	30	4	0	215	12.4%
13	1952–1956	53	1	0	0	54	3.1%
14	1947–1951	17	0	0	0	17	1.0%
15	≦1946	1	0	0	0	1	0.1%
	Total No. of Buildings (Per Unit Group)	800	647	238	43	1,728	
	% in Eastern (Per Unit Groups)	46.3%	37.4%	13.8%	2.5%		100.0%
	Total No. of Buildings under MBIS Year 2014 and % in terms of no. of buildings in 2014	684	366	108	18	1,176	Ranked No. 6 (2014)
		39.6%	21.2%	6.2%	1.0%	68.0%	
	Total No. of Buildings under MBIS by Year 2024 and % in terms of no. of buildings in 2014	766	586	176	31	1,559	Ranked No. 6 (2024)
		44.3%	33.9%	10.2%	1.8%	90.2%	
	Private Buildings under MBIS (Movement 2014–2024)	82	220	68	13	383	Ranking remains unchanged
		4.7%	12.7%	4.0%	0.7%	22.2%	

Note: For Private Buildings Non-Domestic and Domestic Buildings Over 3 storeys

The Eastern District, with 1,728 private buildings ranked sixth in 2014 in terms of number of private buildings. As of 31 December 2014, there was a total of 1,176 (68.0%) private buildings in the district falling under the MBIS also ranked sixth among the 18 districts. In this aspect, in the next 10 years, the projected number will increase to 1,559 (90.2%), and the ranking shall remain unchanged among the districts.

In 1946, there was only one private building in the district. In the period from 1957 to 1991, 1,412 buildings (82.0%) were completed. In terms of size distribution (number of building units) of private buildings in the district, two characteristics are observed. First, the proportion of small-sized buildings is relatively low. As compared with the one of 64.5% in the territory, the proportion of small-sized private buildings in the district was only 46.3%. Second and conversely, the proportion of medium-sized private buildings was relatively higher. As compared to the one of 23.6% in the territory, the proportion of medium-sized private buildings in the district was 37.4%.

As of 31 December 2014, out of the 1,176 (80%) private buildings falling under the MBIS, 684 (39.6%) were small-sized buildings, 366 (21.2%) medium-sized buildings, 108 (6.2%) large-sized buildings and 18 (1.0%) extra-large-sized buildings. In this aspect, in the next 10 years, it is projected that the number shall increase to 1,559 (90.2%). This represents a net increase of 383 (22%) private buildings. The distribution of aged private buildings falling under the MBIS shall be 766 (44.3%) small-sized buildings, 586 (33.9%) medium-sized buildings, 176 (10.2%) large-sized buildings and 31 (1.8%) extra-large-sized buildings. The ranking, in terms of ageing rate (22.2%) in of aged private buildings from 2014 to 2024, shall be the eleventh.

Kwun Tong District, Kwun Tong Road

5.2.7 Kwun Tong District

Key Features

(a) Ranked seventh in 2014 in number of private buildings with a total of 1,026 private buildings in the district

(b) Ranked seventh in 2014 with 677 buildings (66.0%) falling under the MBIS

(c) Shall rank seventh in 2024 with 926 private buildings (90.3%) falling under the MBIS

(d) Changes in 10 years' time

- Ranking shall remain unchanged in terms of number of private buildings falling under the MBIS

- Projected number of buildings under the MBIS shall increase by 249

- Shall rank 10th in terms of the ageing rate (24.3%) of private buildings under the MBIS

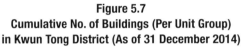

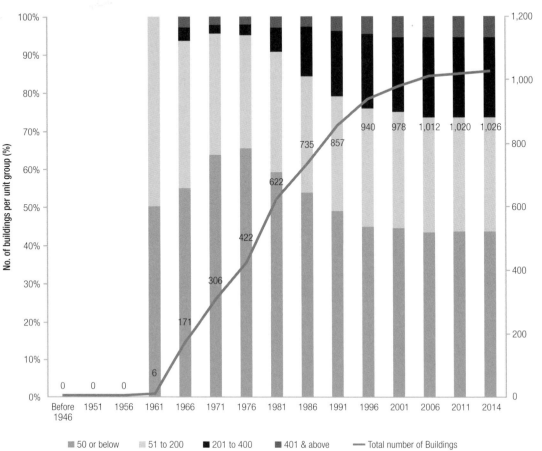

Figure 5.7
Cumulative No. of Buildings (Per Unit Group)
in Kwun Tong District (As of 31 December 2014)

Note: For Non-Domestic Private Buildings and Domestic Buildings over 3 Storeys

Comment

The Kwun Tong District Council (District Council, 2014) highlights that "Kwun Tong is one of the earliest developed urban areas in Hong Kong. Its population has been growing rapidly, and the demand for housing, medical and educational facilities and services has been increasing. In view of this, a number of community development projects, such as the redevelopment of old housing estates and the construction of major parks, have been implemented in recent years".

Table 5.7
Number of Private Buildings (Unit Distribution Per Year Group)
in Kwun Tong District (As of 31 December 2014)

Year of Completion		No. of Private Buildings (Per Unit Groups)					Percentage (%)
		50 or above	51 or 200	201 to 400	401 and above		
		Small	Medium	Large	Extra Large	Total	
1	2012–2014	3	2	1	0	6	
2	2007–2011	4	1	2	1	8	0.8%
3	2002–2006	5	5	24	0	34	3.3%
4	1997–2001	12	8	7	11	28	3.7%
5	1992–1996	2	34	37	10	83	8.1%
6	1987–1991	24	33	51	14	122	12.0%
7	1982–1986	27	29	55	2	113	11.1%
8	1977–1981	92	71	28	9	200	19.6%
9	1972–1976	81	28	5	2	116	11.4%
10	1967–1971	101	31	1	2	135	13.2%
11	1962–1966	91	63	6	5	165	16.2%
12	1957–1961	3	3	0	0	6	0.6%
13	1952–1956	0	0	0	0	0	0.0%
14	1947–1951	0	0	0	0	0	0.0%
15	≦1946	0	0	0	0	0	0.0%
Total No. of Buildings (Per Unit Group)		445	308	217	56	1,026	
% in Kwun Tong (Per Unit Groups)		43.4%	30.0%	21.2%	5.5%		100.0%
Total No. of Buildings under MBIS Year 2014 and % in terms of no. of buildings in 2014		381	210	67	19	677	Ranked No. 7 (2014)
		37.1%	20.5%	6.6%	1.8%	66.0%	
Total No. of Buildings under MBIS by Year 2024 and % in terms of no. of buildings in 2014		433	283	170	40	926	Ranked No. 7 (2024)
		42.2%	27.6%	16.5%	3.9%	90.3%	
Private Buildings under MBIS (Movement 2014–2024)		52	73	103	21	249	Ranking remains unchanged
		5.1%	7.2%	10.0%	2.1%	24.3%	

Note: For Private Buildings Non-Domestic and Domestic Buildings over 3 storeys

The Kwun Tong District, with 1,026 private buildings, ranked seventh in 2014 in terms of number of private buildings among the 18 districts. As of 31 December 2014, there were 677 private buildings (66%) in the district that fall under the MBIS, ranked seventh among the 18 districts. In this aspect, in the next 10 years, it is projected that the number shall increase to 926 (90.3%). The ranking in terms of number of private buildings under the MBIS shall remain unchanged.

Before 1956, record reveals the district had no private buildings. Between 1957 and 1961, there were only three private buildings in the district. From 1962 onwards, the district had drastically developed. Up to 1991, 851 private buildings (83.4%) were completed. In addition, two main points are worth making on the size distribution (number of building units) of private building in this district. First, the proportion of small-sized private buildings is relatively low. Compared with the one of 64.5% in the territory, the proportion of small-sized private buildings in the district was only 43.4%. Second and conversely, the proportions of large-sized and extra-large-sized private buildings was much higher. As compared to the ones of 10.2% and 1.7% in the territory, the proportion of private buildings in the district for large-sized and extra-large-sized private buildings was 21.2% and 5.5% respectively.

As of 31 December 2014, out of the 677 private buildings (66%) falling under the MBIS, 381 were small-sized buildings (37.1%), 210 medium-sized buildings (20.5%), 67 large-sized buildings (6.6%) and 19 extra-large-sized buildings (1.8%). In this aspect, by 2024, it is projected that the number of aged private buildings falling under the MBIS shall increase to 926 (90.3%). This represents a net increase of 249 private buildings (24.3%). The distribution of aged private buildings falling under the MBIS shall be 433 small-sized buildings (42.2%), 283 medium-sized buildings (27.6%), 170 large-sized buildings (16.5%) and 40 extra-large-sized buildings (3.9%). The ranking, in terms of ageing rate (24.3%) of aged private buildings from 2014 to 2024, shall be the 10th.

Southern District, Aberdeen Main Road

5.2.8 Southern District

Key Features

(a) Ranked eighth in 2014 in number of private buildings with a total of 1,004 private buildings in the district

(b) Ranked eighth in 2014 with 620 private buildings (61.8%) falling under the MBIS

(c) Shall rank eighth in 2024 with 819 private buildings (81.6%) falling under the MBIS

(d) Changes in 10 years' time

• Ranking shall remain unchanged in terms of number of private buildings falling under the MBIS

• Projected number of buildings under the MBIS shall increase by 199

• Shall rank twentieth in terms of ageing rate (19.8%) of private buildings under the MBIS

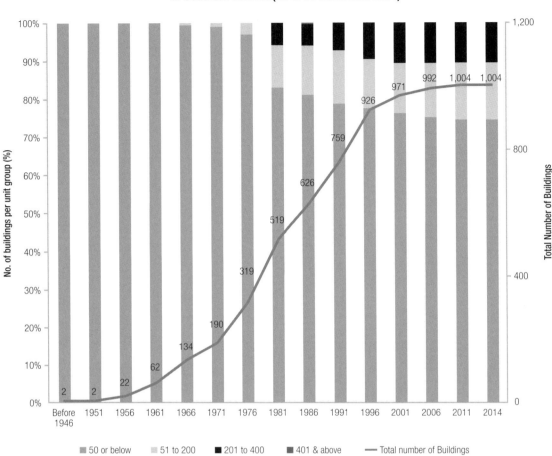

Figure 5.8
Cumulative No. of Buildings (Per Unit Group)
in Southern District (As of 31 December 2014)

Note: For Non-Domestic Private Buildings and Domestic Buildings over 3 Storeys

Comment

The Southern District Council (District Council, 2014) highlights that "around half of the population in the Southern District lives in public rental housing units. Large-scale public housing estates in the district include Wah Fu Estate, Shek Pai Wan Estate and Ap Lei Chau Estate. The rest of the population resides in private homes. Large-scale private residential developments include Baguio Villa, Chi Fu Fa Yuen, Aberdeen Centre, South Horizons, Bel-Air Residence, Redhill Peninsula and Hong Kong Parkview. Despite that the Southern District has been developed into a modern industrial/commercial and residential district, most of its natural landscape, such as the scenic beaches and country parks, remain intact. Moreover, in the vicinity of Aberdeen and Stanley, some residents

Table 5.8
Number of Private Buildings (Unit Distribution Per Year Group)
in Southern District (As of 31 December 2014)

Year of Completion		No. of Private Buildings (Per Unit Groups)					Percentage (%)
		50 or above	51 or 200	201 to 400	401 and above		
		Small	Medium	Large	Extra Large	Total	
1	2012–2014	0	0	0	0	0	0%
2	2007–2011	4	8	0	0	12	1.2%
3	2002–2006	3	15	3	0	21	2.1%
4	1997–2001	22	10	13	0	45	4.5%
5	1992–1996	122	11	34	0	167	16.6%
6	1987–1991	90	25	18	0	133	13.2%
7	1982–1986	77	23	6	1	107	10.7%
8	1977–1981	122	47	31	0	200	19.9%
9	1972–1976	121	8	0	0	129	12.8%
10	1967–1971	55	1	0	0	56	5.6%
11	1962–1966	71	1	0	0	72	7.2%
12	1957–1961	40	0	0	0	40	4.0%
13	1952–1956	20	0	0	0	20	2.0%
14	1947–1951	0	0	0	0	0	0.0%
15	≦1946	2	0	0	0	2	0.2%
Total No. of Buildings (Per Unit Group)		749	149	105	1	1,004	
% in Southern (Per Unit Groups)		74.6%	14.8%	10.5%	0.1%		100.0%
Total No. of Buildings under MBIS Year 2014 and % in terms of no. of buildings in 2014		508	74	37	1	620	Ranked No. 8 (2014)
		50.6%	7.4%	3.7%	0.1%	61.8%	
Total No. of Buildings under MBIS by Year 2024 and % in terms of no. of buildings in 2014		641	107	70	1	819	Ranked No. 8 (2024)
		63.8%	10.7	7.0%	0.1%	81.6%	
Private Buildings under MBIS (Movement 2014–2024)		133	33	33	0	199	Ranking remains unchanged
		13.2%	3.3%	3.3%	0.0%	19.8%	

Note: For Private Buildings Non-Domestic and Domestic Buildings over 3 storeys

still keep traditional customs and rituals. The Aberdeen Fish Market and Typhoon Shelter boast the ambience of a fishing harbour".

The Southern District, with 1,004 private buildings ranked eighth in 2014 in terms of number of private buildings. As of 31 December 2014, there was a total of 620 private buildings (61.8%) in the district that fell under the MBIS and ranked eighth among the 18 districts. In this aspect, in the next 10 years, the projected number shall increase to 819 (81.6%), and the ranking shall remain unchanged in terms of number of private buildings under the MBIS among the 18 districts.

In 1946, there were only two private buildings in the district. In the period from 1972 to 1996, 736 private buildings (73.3%) were completed. In terms of size distribution (number of building units) of private buildings in the district, two characteristics are observed. First, the proportion of small-sized buildings is relatively high. As compared to the one of 64.5% in the territory, the proportion of small-sized private buildings in the district was 74.6%. Second, the proportion of extra-large-sized private buildings was far lower than the city's average. As compared with the average of 1.7% in the territory, there was only one (0.1%) extra-large-sized private building in the district.

As of 31 December 2014, out of the 620 private buildings (61.8%) falling under the MBIS, 508 were small-sized buildings (50.6%), 74 medium-sized buildings (7.4%) and 37 large-sized buildings (3.7%). In this aspect, in the next 10 years, the projected number of aged private buildings falling under the MBIS shall increase to 819 (81.6%). This represents a net increase of 199 private buildings (19.8%). The distribution of aged private buildings falling under the MBIS shall be 641 small-sized buildings (63.8%), 107 medium-sized buildings (10.7%), 70 large-sized buildings (7%) and one extra-large-sized building (0.1%). The ranking, in terms of ageing rate (19.8%) of aged private buildings from 2014 to 2024, shall be the twentieth.

Sha Tin District, Wang Pok Street

5.2.9   Sha Tin District

Key Features

(a)   Ranked ninth in 2014 in number of private buildings with a total of 1003 private buildings in the district

(b)   Ranked 15th in 2014 with 260 buildings (25.9%) falling under the MBIS

(c)   Shall rank 10th in 2024 with 672 private buildings (67%) falling under the MBIS

(d)   Changes in 10 years' time

- Ranking shall move 5 notches up in terms of number of private buildings falling under the MBIS

- Projected number of buildings under the MBIS shall increase by 412

- Shall rank third in terms of ageing rate (41.1%) of private buildings under the MBIS

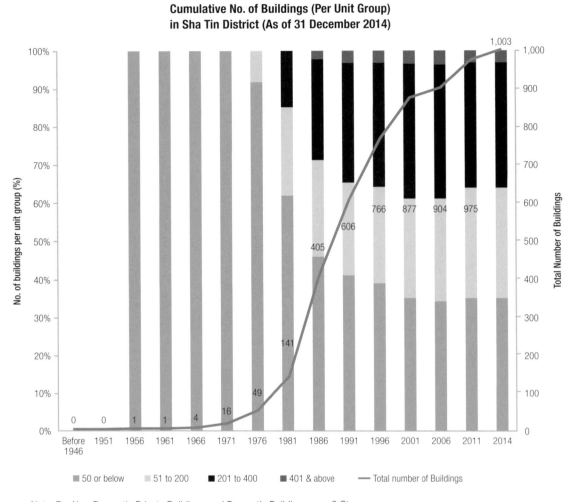

Figure 5.9
Cumulative No. of Buildings (Per Unit Group)
in Sha Tin District (As of 31 December 2014)

Note: For Non-Domestic Private Buildings and Domestic Buildings over 3 Storeys

Comment

The Sha Tin District Council (District Council, 2014) highlights that "Sha Tin is a well developed new town comprising mainly residential areas. About 65% of the population live in public housing which includes rental estates, Tenants Purchase Scheme and Home Ownership Scheme courts. There are also about 29,000 people living in some 48 indigenous villages. In addition, there are four industrial areas for light industries in Sha Tin including Tai Wai, Fo Tan, Siu Lek Yuen and Shek Mun".

The Sha Tin District, with 1003 private buildings ranked ninth in 2014 in terms of number of private buildings. As of 31 December 2014, there was a total of 260 private

Table 5.9
Number of Private Buildings (Unit Distribution Per Year Group)
in Sha Tin District (As of 31 December 2014)

	Year of Completion	No. of Private Buildings (Per Unit Groups)					Percentage (%)
		50 or above	51 or 200	201 to 400	401 and above		
		Small	Medium	Large	Extra Large	Total	
1	2012–2014	20	0	8	0	28	2.79%
2	2007–2011	31	39	1	0	71	7.08%
3	2002–2006	3	14	7	3	27	2.69%
4	1997–2001	8	36	63	4	11	11.07%
5	1992–1996	50	45	60	5	160	15.95%
6	1987–1991	62	45	83	11	201	20.04%
7	1982–1986	98	70	87	9	264	26.32%
8	1977–1981	42	29	21	0	92	9.17%
9	1972–1976	29	4	0	0	33	3.29%
10	1967–1971	12	0	0	0	12	1.20%
11	1962–1966	3	0	0	0	3	0.30%
12	1957–1961	0	0	0	0	0	0.00%
13	1952–1956	1	0	0	0	1	0.10%
14	1947–1951	0	0	0	0	0	0.00%
15	≦1946	0	0	0	0	0	0.00%
	Total No. of Buildings (Per Unit Group)	359	282	330	32	1,003	
	% in Sha Tin (Per Unit Groups)	35.8%	28.1%	32.9%	3.2%		100.0%
	Total No. of Buildings under MBIS Year 2014 and % in terms of no. of buildings in 2014	130	65	61	4	260	Ranked No. 15 (2014)
		12.9%	6.5%	6.1%	0.4%	25.9%	
	Total No. of Buildings under MBIS by Year 2024 and % in terms of no. of buildings in 2014	267	167	216	22	672	Ranked No. 10 (2024)
		26.6%	16.7%	21.6%	2.2%	67.0%	
	Private Buildings under MBIS (Movement 2014–2024)	137	102	155	18	412	Ranking 5 notches up
		13.7%	10.2%	15.5%	1.8%	41.1%	

Note: For Private Buildings Non-Domestic and Domestic Buildings over 3 storeys

buildings (25.9%) in the district that fell under the MBIS, ranking 15th among the 18 districts. In this aspect, in the next 10 years, the projected number will increase to 672 (67%), and the district will soar to rank 10th.

Comparatively speaking, Sha Tin is one of the youngest districts in the territory. On or before 1951, no private buildings were constructed. In the period from 1952 to 1956, there was only one private building in the district. In the period from 1977 to 2001 828 (84.9%) private buildings were completed. In terms of size distribution (number of building units) of private buildings in the district, two characteristics are observed. First, the proportion of small-sized buildings is relatively low. As compared with the average of 64.5% in the territory, the proportion of small-sized buildings in the district was only 35.8%. Second, the proportions of large and extra-large-sized private buildings were relatively higher. As compared to the ones of 10.2% and 1.7% in the territory, the proportions in the district for large-sized and extra-large-sized private buildings were 32.9% and 3.2% respectively.

As of 31 December 2014, out of the 260 private buildings (25.9%) falling under the MBIS, 130 were small-sized buildings (12.9%), 65 medium-sized buildings (6.5%) and 61 large-sized buildings (6.1%). In this aspect, by 2024, it is projected that the number of aged private buildings falling under the MBIS shall drastically increase to 672 (67%). This represents a net increase of 412 private buildings (41.1%). The distribution of aged private buildings falling under the MBIS shall be 267 small-sized buildings (26.6%), 167 medium-sized buildings (16.7%), 216 large-sized buildings (21.6%) and 22 extra-large-sized buildings (2.2%). The ranking, in terms of ageing rate (41.7%) of aged private buildings from 2014 to 2024, shall be the third.

Yuen Long District, Castle Peak Road

5.2.10 Yuen Long District

Key Features

(a) Ranked 10th in 2014 in number of private buildings with a total of 939 private buildings in the district

(b) Ranked 10th in 2014 with 362 buildings (38.6%) falling under the MBIS

(c) Shall rank 12th in 2024 with 601 buildings (64%) falling under the MBIS

(d) Changes in 10 years' time

- Ranking shall move two notch down in terms of number of private buildings falling under the MBIS

- Projected number of buildings under the MBIS shall increase by 239

- Shall rank 9th in terms of ageing rate (25.5%) of private buildings under the MBIS

Figure 5.10
Cumulative No. of Buildings (Per Unit Group)
in Yuen Long District (As of 31 December 2014)

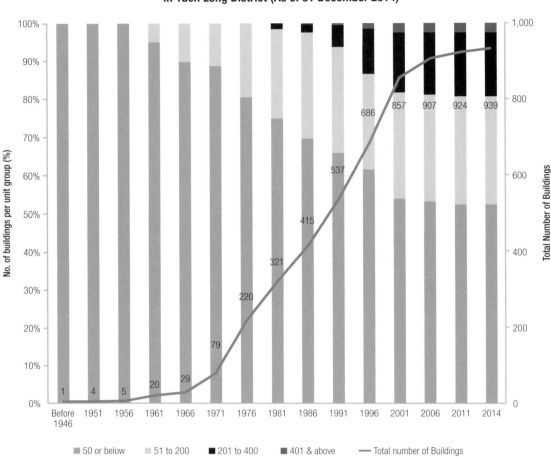

Note: For Non-Domestic Private Buildings and Domestic Buildings over 3 Storeys

Comment

The Yuen Long District Council (District Council, 2014) highlights that "Yuen Long is situated in the northwest of the New Territories in a large alluvial plain surrounded by hills on three sides. The earliest significant settlements in Yuen Long can be traced back to the Sung Dynasty (960–1279 A.D.). The TANG and MAN clans are the earliest known settlers in the district. With the rapid development, Yuen Long district has developed into a town with a population of more than 560,000 residents".

The Yuen Long District, with 939 private buildings ranked 10th in 2014 in terms of number of private buildings among the 18 districts. As of 31 December 2014, there was

Table 5.10
Number of Private Buildings (Unit Distribution Per Year Group)
in Yuen Long District (As of 31 December 2014)

| Year of Completion | | No. of Private Buildings (Per Unit Groups) | | | | | Percentage (%) |
| | | 50 or above | 51 or 200 | 201 to 400 | 401 and above | Total | |
		Small	Medium	Large	Extra Large		
1	2012-2014	0	15	0	0	15	1.60%
2	2007–2011	2	7	8	0	17	1.81%
3	2002–2006	20	16	14	0	50	5.32%
4	1997–2001	40	66	53	12	171	18.21%
5	1992–1996	68	23	50	8	149	15.87%
6	1987–1991	65	33	23	1	122	12.99%
7	1982–1986	49	40	4	1	94	10.01%
8	1977–1981	63	33	5	0	101	10.76%
9	1972–1976	107	34	0	0	141	15.02%
10	1967–1971	44	6	0	0	50	5.32%
11	1962–1966	7	2	0	0	9	0.96%
12	1957–1961	14	1	0	0	15	1.60%
13	1952–1956	1	0	0	0	1	0.11%
14	1947–1951	3	0	0	0	3	0.32%
15	≦1946	1	0	0	0	1	0.11%
Total No. of Building (Per Unit Group)		484	276	157	22	939	
% in Yuen Long (Per Unit Group)		51.5%	29.4%	16.7%	2.3%		100.0%
Total No. of Buildings under MBIS Year 2014 and % in terms of no. of buildings in 2014		261	94	7	0	362	Ranked No. 10 (2014)
		27.8%	10.1%	0.7%	0.1%	38.7%	
Total No. of Buildings under MBIS by Year 2024 and % in terms of no. of buildings in 2014		381	158	56	6	601	Ranked No. 12 (2024)
		40.6%	16.8%	6.0%	0.6%	64.0%	
Private Buildings under MBIS (Movement 2014–2024)		120	63	49	5	239	Ranking 2 notch down
		12.8%	6.7%	5.2%	0.6%	25.4%	

Note: For Private Buildings Non-Domestic and Domestic Buildings over 3 storeys

a total of 362 private buildings in the district (38.7%), aged 30 or above, falling under the MBIS ranked 10th among the 18 districts. In this aspect, in the next 10 years, the projected number will increase to 601 (64%), and the ranking shall move down two notch (i.e., no. 12).

In 1946, there was only one private building (non-domestic and domestic over 3 storeys in height) in the district. In the period from 1947 to 1966, the development of private buildings in the district was very slow. In the period from 1972 to 2001, i.e. 30 years' time, 778 (84.2%) private buildings were completed. It is worth noting that the proportion of small-sized buildings is relatively low. As compared to the one of 64.5% in the territory, the proportion of small-sized private buildings in the district was only 51.5%.

As of 31 December 2014, out of the 362 private buildings (38.7%) falling under the MBIS, 260 were small-sized buildings (27.8%), 94 medium-sized buildings (10.1%) and seven large-sized buildings (0.7%). In this aspect, in the next 10 years, it is projected that the number shall increase to 601 (64%). This represents a net increase of 239 private buildings (25.5%). The distribution of aged private buildings falling under the MBIS shall be 381 small-sized buildings (40.6%), 158 medium-sized buildings (16.8%), 56 large-sized buildings (6%) and six extra-large-sized buildings (0.6%). The ranking, in terms of ageing rate (25.5%) of aged private buildings from 2014 to 2024, shall be the 9th.

Tsuen Wan District, Chuen Lung Street

5.2.11 Tsuen Wan District

Key Features

(a) Ranked 11th in 2014 in number of private buildings with a total of 865 private buildings in the district

(b) Ranked ninth in 2014 with 502 buildings (58%) falling under the MBIS

(c) Shall rank ninth in 2024 with 748 buildings (86.5%) falling under the MBIS

(d) Changes in 10 years' time

- Ranking shall remain unchanged in terms of number of private buildings falling under the MBIS

- Projected number of buildings under the MBIS shall increase by 246

- Shall be ranked seventh in terms of ageing rate (28.4%) of private buildings under the MBIS

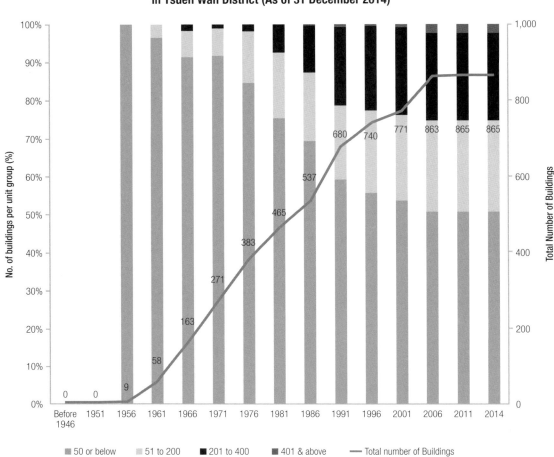

Figure 5.11
Cumulative No. of Buildings (Per Unit Group)
in Tsuen Wan District (As of 31 December 2014)

Note: For Non-Domestic Private Buildings and Domestic Buildings over 3 Storeys

Comment

The Tsuen Wan District Council (District Council, 2014) highlights that "Tsuen Wan was the first new town to be developed in Hong Kong. The industrial and commercial activities in Tsuen Wan grew rapidly early on, though nowadays most of the industrial buildings have been either knocked down or adapted for other purposes in line with the industrial decline of Hong Kong".

The Tsuen Wan District, with 865 private buildings, ranked 11th in 2014 in terms of number of private buildings among the 18 districts. As of 31 December 2014, there was a total of 502 private buildings (58%) in the district, aged 30 years or above, falling under

Table 5.11
Number of Private Buildings (Unit Distribution Per Year Group)
in Tsuen Wan District (As of 31 December 2014)

| | | No. of Private Buildings (Per Unit Groups) | | | | | |
| | | 50 or above | 51 or 200 | 201 to 400 | 401 and above | | Percentage |
	Year of Completion	Small	Medium	Large	Extra Large	Total	(%)
1	2012–2014	0	0	0	0	0	0.0%
1	2007–2011	0	2	0	0	2	0.2%
2	2002–2006	24	33	20	15	92	10.6%
3	1997–2001	2	12	16	1	31	3.6%
4	1992–1996	9	29	22	0	60	6.9%
5	1987–1991	31	35	73	4	143	16.5%
6	1982–1986	22	16	33	1	72	8.3%
7	1977–1981	25	29	28	0	82	9.5%
8	1972–1976	77	31	4	0	112	12.9%
9	1967–1971	99	9	0	0	108	12.5%
10	1962–1966	93	9	3	0	105	12.1%
11	1957–1961	47	2	0	0	49	5.7%
12	1952–1956	9	0	0	0	9	1.0%
13	1947–1951	0	0	0	0	0	0.0%
14	≦1946	0	0	0	0	0	0.0%
	Total No. of Buildings (Per Unit Group)	438	207	199	21	865	
	% in Tsuen Wan (Per Unit Groups)	50.6%	23.9%	23.0%	2.4%		100.0%
	Total No. of Buildings under MBIS Year 2014 and % in terms of no. of buildings in 2014	361	88	52	1	502	Ranked No. 9 (2014)
		41.7%	10.2%	6.0%	0.1%	58.0%	
	Total No. of Buildings under MBIS by Year 2024 and % in terms of no. of buildings in 2014	430	153	160	5	748	Ranked No. 9 (2024)
		49.7%	17.7%	18.5%	0.6%	86.5%	
	Private Buildings under MBIS (Movement 2014–2024)	69	65	108	4	246	Ranking remains unchanged
		8.0%	7.5%	12.5%	0.5%	28.4%	

Note: For Private Buildings Non-Domestic and Domestic Buildings over 3 storeys

the MBIS ranked ninth among the 18 districts. In this aspect, in the next 10 years, the projected number will increase to 748 (86.5%), and the ranking shall remain unchanged.

Comparatively speaking, Tsuen Wan is one of the youngest districts in the territory. On or before 1951, no private buildings were constructed. In the period from 1952 to 1956, there were only nine private buildings in the district. In the period from 1962 to 1991, 622 private buildings (71.9%) were completed. In terms of size distribution (number of building units) of private buildings in the district, two characteristics are observed. First, the proportion of small-sized buildings is relatively low. As compared with the average of 64.5% in the territory, the proportion of small-sized private buildings in the district was 50.6%. Second, the proportion of large-sized private buildings was much higher than that of the territory. As compared with the mean of 10.2% in the territory, the proportion of large-sized private buildings in the district was 23%.

As of 31 December 2014, out of the 502 private buildings (58%) falling under the MBIS, 361 were small-sized buildings (41.7%), 88 medium-sized buildings (10.2%) and 52 large-sized buildings (6%). In this aspect, by 2024, the projected number shall increase to 748 (86.5%). This represents a net increase of 246 private buildings (28.4%). The distribution of aged private buildings falling under the MBIS shall be 430 small-sized buildings (49.7%), 153 medium-sized buildings (17.7%), 160 large-sized buildings (18.5%) and five extra-large-sized buildings (0.6%). The ranking, in terms of ageing rate (28.4%) of aged private buildings from 2014 to 2024, shall be the seventh.

Tai Po District, Kwong Fu Road

5.2.12 Tai Po District

Key Features

    (a)   Ranked 12th in 2014 in number of private buildings with a total of 835 private buildings in the district

    (b)   Ranked 13th in 2014 with 311 buildings (37.2%) falling under the MBIS

    (c)   Shall rank 13th in 2024 with 582 buildings (69.7%) falling under the MBIS

    (d)   Changes in 10 years' time

- Ranking remains unchanged in terms of number of private buildings falling under the MBIS

- Projected number of buildings under the MBIS shall increase by 271

- Shall rank fourth in terms of ageing rate (32.5%) of private buildings under the MBIS

Figure 5.12
Cumulative No. of Buildings (Per Unit Group)
in Tai Po District (As of 31 December 2014)

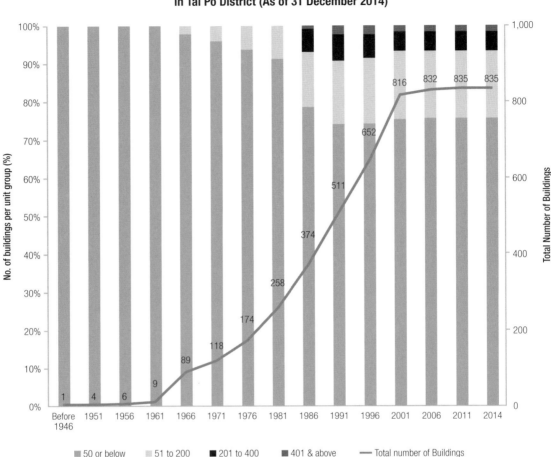

Note: For Non-Domestic Private Buildings and Domestic Buildings over 3 Storeys

Comment

The Tai Po District Council (District Council, 2014) highlights that "Tai Po District is the second largest administrative district in Hong Kong. Tai Po District is a community where the old meets the new, and features a harmonious blend of rural tranquility and urban vibrancy. Upon rapid development over the years, Tai Po District has grown from a simple village to a well-established modern town".

The Tai Po District, with 835 private buildings ranked 12th in 2014 in terms of number of private buildings among the 18 districts. As of 31 December 2014, there was a total of

Table 5.12
Number of Private Buildings (Unit Distribution Per Year Group)
in Tai Po District (As of 31 December 2014)

		No. of Private Buildings (Per Unit Groups)					
		50 or above	51 or 200	201 to 400	401 and above		Percentage (%)
Year of Completion		Small	Medium	Large	Extra Large	Total	
1	2012–2014	0	0	0	0	0	0.0%
1	2007–2011	3	0	0	0	3	0.4%
2	2002–2006	15	1	0	0	16	1.9%
3	1997–2001	130	34	0	0	164	19.6%
4	1992–1996	105	27	7	2	141	16.9%
5	1987–1991	85	30	12	10	137	16.4%
6	1982–1986	58	32	23	3	116	13.9%
7	1977–1981	72	12	0	0	84	10.1%
8	1972–1976	50	6	0	0	56	6.7%
9	1967–1971	26	3	0	0	29	3.5%
10	1962–1966	78	2	0	0	80	9.6%
11	1957–1961	3	0	0	0	3	0.4%
12	1952–1956	2	0	0	0	2	0.2%
13	1947–1951	2	0	0	0	3	0.4%
14	≦1946	1	0	0	0	1	0.1%
Total No. of Buildings (Per Unit Group)		631	147	42	15	835	
% in Tai Po (Per Unit Groups)		75.6%	17.6%	5.0%	1.8%		100.0%
Total No. of Buildings under MBIS Year 2014 and % in terms of no. of buildings in 2014		261	38	11	1	311	Ranked No. 13 (2014)
		31.3%	4.6%	1.3%	0.1%	37.2%	
Total No. of Buildings under MBIS by Year 2024 and % in terms of no. of buildings in 2014		430	99	39	14	582	Ranked No. 13 (2024)
		51.5%	11.9%	4.7%	1.7%	69.7%	
Private Buildings under MBIS (Movement 2014–2024)		169	61	28	13	271	Ranking remains unchanged
		20.2%	7.3%	3.4%	1.5%	32.5%	

Note: For Private Buildings Non–Domestic and Domestic Buildings over 3 storeys

311 private buildings (37.2%) in the district, aged 30 or above, falling under the MBIS was ranked 14th among the 18 districts. In this aspect, in the next 10 years, the projected number will increase to 582 (69.7%) and the ranking shall remain unchanged.

Comparatively speaking, Tai Po is one of the youngest districts in the territory. In 1946, there was only one private building in the district. During the period from 1947 to 1961, the development of private buildings was very slow, with only 8 private buildings were completed. In the period from 1977 to 2001, 642 private buildings (76.9%) were completed. In terms of size distribution (number of building units) of private buildings in the district, two characteristics are observed. First, the proportion of small-sized buildings is relatively high. As compared with the average of 64.5% in the territory, the proportion of small-sized private buildings in the district was 75.6%. Second and conversely, the proportion of large-sized private buildings was lower than that of the territory. As compared to the average of 10.2% in the territory, the proportion of large-sized private buildings in the district was only 5.0%.

As of 31 December 2014, out of the 311 private buildings (37.2%) falling under the MBIS, 261 were small-sized buildings (31.3%) and 38 were medium-sized buildings (4.6%). In this aspect, in the next 10 years, it is projected that the total number shall increase to 582 (69.7%). This represents a net increase of 271 private buildings (32.5%). The distribution of aged private buildings falling under the MBIS shall be 430 small-sized buildings (51.5%), 99 medium-sized buildings (11.9%), 39 large-sized buildings (4.7%) and 14 extra-large-sized buildings (1.7%). The ranking, in terms of ageing rate (32.5%) of aged private buildings from 2014 to 2024, shall be the fourth.

Tuen Mun District, Yan Ching Street

5.2.13 Tuen Mun District

Key Features

(a) Ranked 13th in 2014 in number of private buildings with a total of 784 private buildings in the district

(b) Ranked 16th in 2014 with 254 buildings (32.4%) falling under the MBIS

(c) Shall rank 11th in 2024 with 657 buildings (83.8%) falling under the MBIS

(d) Changes in 10 years' time

- Ranking shall move five notches up in terms of number of private buildings falling under the MBIS

- Projected number of buildings under the MBIS shall increase by 403

- Shall rank first in terms of ageing rate (51.4%) of private buildings under the MBIS

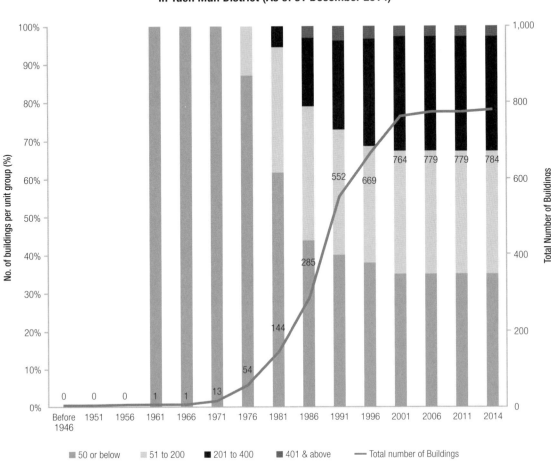

Figure 5.13
Cumulative No. of Buildings (Per Unit Group)
in Tuen Mun District (As of 31 December 2014)

Note: For Non-Domestic Private Buildings and Domestic Buildings over 3 Storeys

Comment

The Tuen Mun District Council (District Council, 2014) highlights that "it is one of the new towns first developed in Hong Kong. The development of Tuen Mun New Town started in the 1960's when its population was a mere of 20,000 odd. With extensive reclamation and the completion of more and more high rise buildings, it has grown into a big town with over 500,000 inhabitants. Besides village houses indigenous to the rural area, there are also a variety of other residential buildings in the district of Tuen Mun, including newly built village houses, private housing estates, single block buildings, low density residences, public housing and Home Ownership Scheme estates provided by the Government".

Table 5.13
Number of Private Buildings (Unit Distribution Per Year Group)
in Tuen Mun District (As of 31 December 2014)

	Year of Completion	No. of Private Buildings (Per Unit Groups)					
		50 or above	51 or 200	201 to 400	401 and above		Percentage (%)
		Small	Medium	Large	Extra Large	Total	
1	2012–2014	1	0	4	0	5	0.6%
2	2007–2011	0	0	0	0	0	0.0%
3	2002–2006	6	4	5	0	15	1.9%
4	1997–2001	13	44	38	0	95	12.1%
5	1992–1996	33	23	60	1	117	14.9%
6	1987–1991	96	81	78	12	267	34.1%
7	1982–1986	36	53	43	9	141	18.0%
8	1977–1981	42	40	8	0	90	11.5%
9	1972–1976	34	7	0	0	41	5.2%
10	1967–1971	12	0	0	0	12	1.5%
11	1962–1966	0	0	0	0	0	0.0%
12	1957–1961	1	0	0	0	1	0.1%
13	1952–1956	0	0	0	0	0	0.0%
14	1947–1951	0	0	0	0	0	0.0%
15	≦1946	0	0	0	0	0	0.0%
	Total No. of Buildings (Per Unit Group)	274	252	236	22	784	
	% in Tuen Mun (Per Unit Groups)	34.9%	32.1%	30.1%	2.8%		100.0%
	Total No. of Buildings under MBIS Year 2014 and % in terms of no. of buildings in 2014	127	87	35	5	254	Ranked No. 16 (2014)
		16.2%	11.1%	4.5%	0.7%	32.4%	
	Total No. of Buildings under MBIS by Year 2024 and % in terms of no. of buildings in 2014	256	207	171	23	657	Ranked No. 11 (2024)
		32.6%	26.4%	21.8%	3.0%	83.8%	
	Private Buildings under MBIS (Movement 2014–2024)	129	120	136	18	403	Ranking 5 notches up
		16.4%	15.3%	17.4%	2.3%	51.4%	

Note: For Private Buildings Non–Domestic and Domestic Buildings over 3 storeys

The Tuen Mun District, with 784 private buildings, ranked 13th in 2014 in terms of number of private buildings among the 18 districts. As of 31 December 2014, there was a total of 254 private buildings (32.4%) in the district, aged 30 or above, falling under the MBIS ranked 16th among the 18 districts. In this aspect, in the next 10 years, the projected number will increase to 657 (83.8%) and the district shall soar to rank 11th.

Comparatively speaking, Tuen Mun is also one of the youngest districts in the territory. On or before 1956, no private buildings were constructed. In 1961, there was only one private building. In the period from 1977 to 2001, 710 (91.1%) private buildings were completed. In terms of size distribution (number of building units) of private buildings in the district, two characteristics are observed. First, the proportion of small-sized buildings is relatively low. As compared with the average of 64.5% in the territory, the proportion of small-sized private buildings in the district was only 34.9%. Second and conversely, the proportion of large-sized private buildings was higher than that of the territory. As compared with the average of 10.2% in the territory, the proportion of large-sized private buildings in the district was 30.1%.

As of 31 December 2014, out of the 254 private buildings (32.4%) falling under the MBIS, 127 were small-sized buildings (16.2%), 87 medium-sized buildings (11.1%) and 35 large-sized private buildings (4.5%). By 2024, the number of aged private buildings falling under the MBIS shall increase to 657 (83.8%), representing a net increase of 403 private buildings (51.4%). The distribution of aged private buildings falling under the MBIS shall be 256 small-sized buildings (32.6%), 207 medium-sized buildings (26.4%), 171 large-sized buildings (21.8%) and 23 extra-large-sized buildings (3.0%). The ranking, in terms of ageing rate (51.4%) of aged private buildings from 2014 to 2024, shall be the first.

North District, Fanling, Luen Hing Street

5.2.14 North District

Key Features

(a) Ranked 14th in 2014 in number of private buildings with a total of 710 private buildings in the district

(b) Ranked 14th in 2014 with 269 buildings (37.9%) falling under the MBIS

(c) Shall rank 15th in 2024 with 473 buildings (66.6%) falling under the MBIS

(d) Changes in 10 years' time

   • Ranking shall move one notch down in terms of number of private buildings falling under the MBIS

   • Projected number of buildings under the MBIS shall increase by 204

   • Shall rank sixth in terms of ageing rate (28.7%) of private buildings under the MBIS

Figure 5.14
Cumulative No. of Buildings (Per Unit Group)
in North District (As of 31 December 2014)

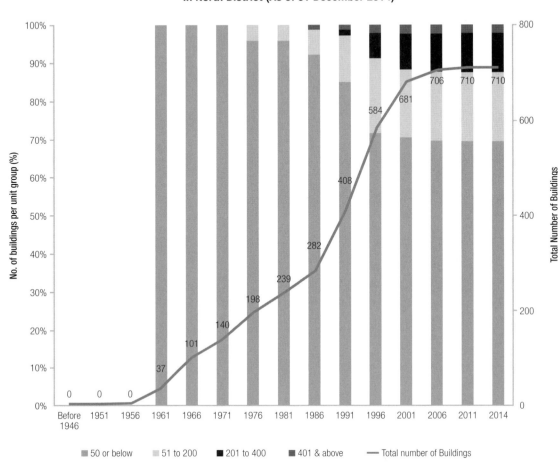

Note: For Non-Domestic Private Buildings and Domestic Buildings over 3 Storeys

Comment

The North District Council (District Council, 2014) highlights that "North District has a long history where the traditional clan culture of 'walled village' is particularly prominent and well preserved. There are 117 villages in North District. Many villagers of same surnames and clans still live there. The four rural committees, namely Sheung Shui, Fanling, Ta Kwu Ling and Sha Tau Kok District Rural Committees, help to manage rural affairs and liaise with the villagers. North District is one of the districts where traditional rural features of the New Territories are best preserved".

The North District, with 710 private buildings, ranked 14th in 2014 in terms of number of private buildings among the 18 districts. As of 31 December 2014, there was a total of

Table 5.14
Number of Private Buildings (Unit Distribution Per Year Group)
in North District (As of 31 December 2014)

| | Year of Completion | No. of Private Buildings (Per Unit Groups) | | | | | Percentage (%) |
| | | 50 or above | 51 or 200 | 201 to 400 | 401 and above | Total | |
		Small	Medium	Large	Extra Large		
1	2012–2014	0	0	0	0	0	0.0%
2	2007–2011	2	0	2	0	4	0.6%
3	2002–2006	10	9	6	0	25	3.5%
4	1997–2001	62	5	27	3	97	13.7%
5	1992–1996	72	64	32	8	176	24.8%
6	1987–1991	87	32	6	1	126	17.7%
7	1982–1986	31	8	0	4	43	6.1%
8	1977–1981	39	2	0	0	31	5.8%
9	1972–1976	50	8	0	0	58	8.2%
10	1967–1971	39	0	0	0	39	5.5%
11	1962–1966	64	0	0	0	64	9.0%
12	1957–1961	37	0	0	0	37	5.2%
13	1952–1956	0	0	0	0	0	0.0%
14	1947–1951	0	0	0	0	0	0.0%
15	≦1946	0	0	0	0	0	0.0%
Total No. of Buildings (Per Unit Group)		493	128	73	16	710	
% in North (Per Unit Groups)		69.4%	18.0%	10.3%	2.3%		100.0%
Total No. of Buildings under MBIS Year 2014 and % in terms of no. of buildings in 2014		253	14	0	2	269	Ranked No. 14 (2014)
		35.6%	2.0%	0.0%	0.3%	37.9%	
Total No. of Buildings under MBIS by Year 2024 and % in terms of no. of buildings in 2014		365	78	21	9	473	Ranked No. 15 (2024)
		51.4%	11.0%	3.0%	1.2%	66.6%	
Private Buildings under MBIS (Movement 2014–2024)		112	64	21	7	204	Ranking 1 notch up
		15.8%	9.0%	3.0%	0.9%	28.7%	

Note: For Private Buildings Non–Domestic and Domestic Buildings over 3 storeys

269 private buildings in the district, aged 30 or above, falling under the MBIS, making the district the 14th among the 18 districts. In this aspect, in the next 10 years, the projected number will increase to 473 (66.6%), and the ranking shall be notched down by one.

On or before 1956, record reveals that no private buildings were constructed. In the period from 1987 to 2001, 399 private buildings (56.2%) were completed. In terms of size distribution (number of building units) of private buildings in the district, only small-sized private buildings were completed during the period from 1957 to 1971.

As of 31 December 2014, out of the 269 private buildings (37.9%) falling under the MBIS, 253 were small-sized buildings (35.6%) and 14 medium-sized buildings (2%). In this aspect, in the next 10 years, the number shall increase to 473 (66.6%). This represents a net increase of 204 private buildings (28.7%). The distributions of aged private buildings falling under the MBIS shall be 365 small-sized buildings (51.4%), 78 medium-sized buildings (11%), 21 large-sized buildings (3%) and 9 extra-large-sized buildings (1.2%). The ranking, in terms of ageing rate (28.7%) of aged private buildings from 2014 to 2024, shall be the ninth.

Kwai Tsing District, Kwai Foo Road

5.2.15 Kwai Tsing District

Key Features

(a) Ranked 15th in 2014 in number of private buildings with a total of 639 private buildings in the district

(b) Ranked 11th in 2014 with 361 buildings (56.5%) falling under the MBIS

(c) Shall rank 14th in 2024 with 555 buildings (86.9%) falling under the MBIS

(d) Changes in 10 years' time

- Ranking shall move three notch down in terms of number of private buildings falling under the MBIS

- Projected number of buildings under the MBIS shall increase by 194

- Shall rank fifth in terms of ageing rate (30.4%) of private buildings under the MBIS

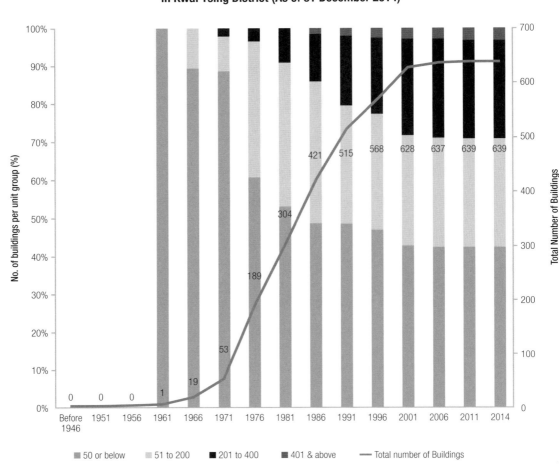

Figure 5.15
Cumulative No. of Buildings (Per Unit Group)
in Kwai Tsing District (As of 31 December 2014)

Note: For Non-Domestic Private Buildings and Domestic Buildings over 3 Storeys

Comment

The Kwai Tsing District Council (District Council, 2014) highlights that "there are 24 public housing estates, 15 Home Ownership Scheme estates, three Sandwich Class Housing Scheme estates and a number of private housing estates in Kwai Tsing. With some large-scale shopping centres and a comprehensive transport network including the Tsing Ma Bridge, Route 3, Tsing Yi North Coastal Road, MTR and Airport Railway, the residents in the District can enjoy a comfortable and convenient living environment".

The Kwai Tsing District, with 639 private buildings, ranked 15th in 2014 in terms of number of private buildings among the 18 districts. As of 31 December 2014, there was a

Table 5.15
Number of Private Buildings (Unit Distribution Per Year Group)
in Kwai Tsing District (As of 31 December 2014)

Year of Completion		No. of Private Buildings (Per Unit Groups)					Percentage (%)
		50 or above	51 or 200	201 to 400	401 and above		
		Small	Medium	Larg	Extra Large	Total	
1	2012–2014	0	0	0	0	0	0.0%
2	2007–2011	0	0	0	2	2	0.3%
3	2002–2006	1	1	7	0	9	1.4%
4	1997–2001	3	8	45	4	60	9.4%
5	1992–1996	16	14	19	4	53	8.3%
6	1987–1991	45	3	42	4	94	14.7%
7	1982–1986	44	41	26	6	117	18.3%
8	1977–1981	47	47	21	0	115	18.0%
9	1972–1976	68	63	5	0	136	21.3%
10	1967–1971	30	3	1	0	34	5.3%
11	1962–1966	16	2	0	0	18	2.8%
12	1957–1961	1	0	0	0	1	0.2%
13	1952–1956	0	0	0	0	0	0.0%
14	1947–1951	0	0	0	0	0	0.0%
15	≦1946	0	0	0	0	0	0.0%
Total No. of Buildings (Per Unit Group)		271	182	166	20	639	
% in Kwai Tsing (Per Unit Groups)		42.4%	28.5%	26.0%	3.1%		100.0%
Total No. of Buildings under MBIS Year 2014 and % in terms of no. of buildings in 2014		183	135	40	3	361	Ranked No. 11 (2014)
		28.7%	21.1%	6.2%	0.5%	56.5%	
Total No. of Buildings under MBIS by Year 2024 and % in terms of no. of buildings in 2014		266	170	107	12	555	Ranked No. 14 (2024)
		41.6%	26.6%	16.8%	1.9%	86.9%	
Private Buildings under MBIS (Movement 2014–2024)		83	35	67	9	194	Ranking 3 notches down
		13.0%	5.5%	10.5%	1.5%	30.4%	

Note: For Private Buildings Non-Domestic and Domestic Buildings over 3 storeys

total of 361 private buildings (56.5%) in the district, aged 30 or above, falling under the MBIS, making the district the 11th among the 18 districts in Hong Kong. In this aspect, in the next 10 years, the projected number shall increase to 555 (56.9%), and the ranking shall be notched down by three.

In 1956, no private buildings were constructed. In the period from 1957 to 1961, only one private building was completed. In the period from 1972 to 1991, 462 private buildings were completed. In terms of size distribution (number of building units) of private buildings in the district, two characteristics are observed. First, the proportion of small-sized buildings is relatively low. As compared with the average of 64.5% in the territory, the proportion of small-sized private buildings in the district was only 42.4%. Second, the proportions of large-sized and extra-large-sized private buildings were higher than that of the territory. As compared with the averages of 10.2% and 1.7% in the territory, the proportions of large-sized and extra-large-sized private buildings in the district were 26.0% and 3.1% respectively.

As of 31 December 2014, out of the 361 (56.5%) private buildings falling under the MBIS, 183 were small-sized buildings (28.7%), 135 medium-sized buildings (21.1%) and 40 large-sized buildings (6.2%). In this aspect, by 2024, the projected number shall increase to 555 (86.9%). This represents a net increase of 194 private buildings (30.4%). The distribution of aged private buildings falling under the MBIS shall be 266 small-sized buildings (41.6%), 170 medium-sized buildings (26.6%), 107 large-sized buildings (16.8%) and 12 extra-large-sized buildings (1.9%). The ranking, in terms of ageing rate (30.4%) of aged private buildings from 2014 to 2024, shall be the fifth.

Wong Tai Sin District, Fung Tak Road

5.2.16 Wong Tai Sin District

Key Features

(a) Ranked 16th in 2014 in number of private buildings with a total of 497 private buildings in the district

(b) Ranked 12th in 2014 with 340 buildings (68.4%) falling under the MBIS

(c) Shall rank 16th in 2024 with 414 buildings (83.3%) falling under the MBIS

(d) Changes in 10 years' time

  • Ranking shall move four notches down in terms of number of private buildings falling under the MBIS

  • Projected number of buildings under the MBIS shall increase by 74

  • Shall rank 15th in terms of ageing rate (14.9%) of private buildings under the MBIS

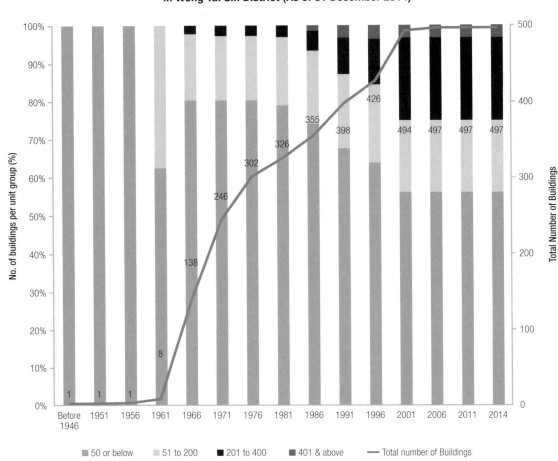

Figure 5.16
Cumulative No. of Buildings (Per Unit Group)
in Wong Tai Sin District (As of 31 December 2014)

50 or below   51 to 200   201 to 400   401 & above   Total number of Buildings

Note: For Non-Domestic Private Buildings and Domestic Buildings over 3 Storeys

Comment

The Wong Tai Sin District Council (District Council, 2014) highlights that "Wong Tai Sin District is characterized by extensive public housing. At present, 85% of the population in the District live in public rental housing, home ownership flats; while the remaining 15% live in various kinds of private housing. Construction of public housing in the District started in 1957 with the first resettlement block of the Mark I type built at Lo Fu Ngam (later renamed as Lok Fu). Over the years, the District has undergone a facelift with the gradual clearance of its squatter areas and redevelopment of its public housing estates".

The Wong Tai Sin District, with 497 private buildings, ranked 16th in 2014 in terms of number of private buildings among the 18 districts. As of 31 December 2014, there was

Table 5.16
Number of Private Buildings (Unit Distribution Per Year Group)
in Wong Tai Sin District (As of 31 December 2014)

Year of Completion		No. of Private Buildings (Per Unit Groups)					Percentage (%)
		50 or above	51 or 200	201 to 400	401 and above		
		Small	Medium	Large	Extra Large	Total	
1	2012–2014	0	0	0	0	0	0.0%
2	2007–2011	0	0	0	0	0	0.0%
3	2002–2006	2	0	1	0	3	0.6%
4	1997–2001	5	6	55	2	68	13.7%
5	1992–1996	3	10	13	2	28	5.6%
6	1987–1991	6	9	20	8	43	8.7%
7	1982–1986	5	11	8	5	29	5.8%
8	1977–1981	15	7	2	0	24	4.8%
9	1972–1976	45	9	2	0	56	11.3%
10	1967–1971	87	18	3	0	108	21.7%
11	1962–1966	106	21	3	0	130	26.2%
12	1957–1961	4	3	0	0	7	1.4%
13	1952–1956	0	0	0	0	0	0.0%
14	1947–1951	0	0	0	0	0	0.0%
15	≦1946	1	0	0	0	1	0.2%
Total No. of Building (Per Unit Group)		279	94	107	17	497	
% in Wai Tai Sin (Per Unit Groups)		56.1%	18.9%	21.5%	3.4%		100.0%
Total No. of Buildings under MBIS Year 2014 and % in terms of no. of buildings in 2014		261	63	14	2	340	Ranked No. 12 (2014)
		52.5%	12.7%	2.8%	0.5%	68.4%	
Total No. of Buildings under MBIS by Year 2024 and % in terms of no. of buildings in 2014		272	83	45	14	414	Ranked No. 16 (2024)
		54.7%	16.7%	9.1%	2.8%	83.3%	
Private Buildings under MBIS (Movement 2014–2024)		11	20	31	12	74	Ranking 4 notches down
		2.2%	4.0%	6.2%	2.4%	14.9%	

Note: For Private Buildings Non–Domestic and Domestic Buildings over 3 storeys

a total of 340 private buildings (68.4%) in the district, aged 30 or above, falling under the MBIS, making the district the 12th among the 18 districts. In this aspect, in the next 10 years, the projected number shall increase to 414 (83.3%) and the ranking shall be notched down by four.

In 1946, there was only one private building constructed in the district. In the period from 1947 to 1956, no private buildings were completed. In the period from 1962 to 1976, contrastingly, 294 private buildings (59.2%) were completed. In terms of size distribution (number of building units) of private buildings in the district, two characteristics are observed. First, the proportion of small-sized buildings is relatively low. As compared with the average of 64.5% in the territory, the proportion of small-sized private building in the district was only 56.1%. Second and conversely, the proportion of large-sized private buildings was higher than that of the territory. As compared with the averages of 10.2% and 1.7% in the territory, the proportions of large-sized and extra-large-sized private buildings in the district were 21.5% and 3.4% respectively.

As of 31 December 2014, out of the 340 private buildings (68.4%) falling under the MBIS, 261 were small-sized buildings (52.5%), 63 medium-sized buildings (12.7%) and 14 large-sized buildings (2.8%). In this aspect, in the next 10 years, the projected number will increase to 414 (82.3%). This represents a net increase of 74 (14.9%) private buildings. As of 31 December 2024, the distributions of aged private buildings falling under the MBIS shall be 272 small-sized buildings (54.7%), 83 medium-sized buildings (16.7%), 45 large-sized buildings (9.1%) and 14 extra-large-sized buildings (2.8%). The ranking, in terms of ageing rate (14.9%) of aged private buildings from 2014 to 2024, shall be the 15th.

Sai Kung District, Marina Cove

5.2.17 Sai Kung District

Key Features

(a) Ranked 17th in 2014 in number of private buildings with a total of 480 private buildings in the district

(b) Ranked 18th in 2014 with 55 buildings (11.5%) falling under the MBIS

(c) Shall rank 18th in 2024 with 179 buildings (37.3%) falling under the MBIS

(d) Changes in 10 years' time

- Ranking remains unchanged in terms of number of private buildings falling under the MBIS

- Projected number of buildings under the MBIS shall increase by 124

- Shall rank 8th in terms of ageing rate (25.8%) of private buildings under the MBIS

Figure 5.17
Cumulative No. of Buildings (Per Unit Group)
in Sai Kung District (As of 31 December 2014)

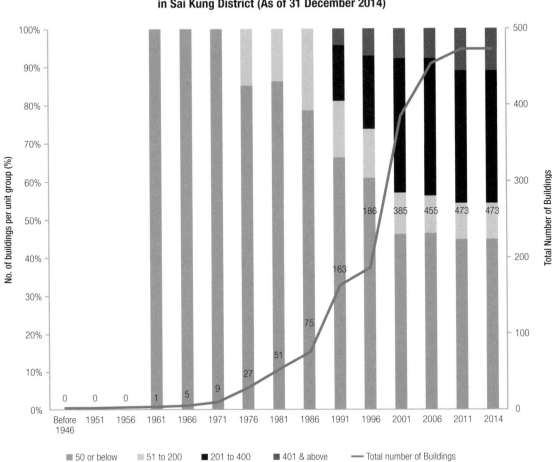

Note: For Non-Domestic Private Buildings and Domestic Buildings over 3 Storeys

Comment

The Sai Kung District Council (District Council, 2014) highlights that "it is the second largest administrative district in Hong Kong in area. It is also growing in popularity in recent years as a seafood centre. As for Tseung Kwan O, it is one of Hong Kong's latest and most rapidly developing new towns".

The Sai Kung District, with 480 private buildings, ranked 17th in 2014 in terms of number of private buildings among the 18 districts. As of 31 December 2014, there was a total of 55 private buildings (11.5%) in the district, aged 30 or above, falling under the

Table 5.17
Number of Private Buildings (Unit Distribution Per Year Group)
in Sai Kung District (As of 31 December 2014)

	Year of Completion	No. of Private Buildings (Per Unit Groups)					Percentage (%)
		50 or above	51 or 200	201 to 400	401 and above		
		Small	Medium	Large	Extra Large	Total	
1	2012–2014	0	7	0	0	7	1.46%
2	2007–2011	1	0	2	15	18	3.75%
3	2002–2006	34	2	27	7	70	14.58%
4	1997–2001	64	18	100	17	199	41.46%
5	1992–1996	5	0	12	6	23	4.79%
6	1987–1991	49	8	24	7	88	18.33%
7	1982–1986	15	9	0	0	24	5.00%
8	1977–1981	21	3	0	0	24	5.00%
9	1972–1976	14	4	0	0	18	3.75%
10	1967–1971	4	0	0	0	4	0.83%
11	1962–1966	4	0	0	0	4	0.83%
12	1957–1961	1	0	0	0	1	0.21%
13	1952–1956	0	0	0	0	0	0.0%
14	1947–1951	0	0	0	0	0	0.0%
15	≦1946	0	0	0	0	0	0.0%
	Total No. of Buildings (Per Unit Group)	212	51	165	52	480	
	% in Sai Kung (Per Unit Groups)	44.2%	10.6%	34.4%	10.8%		100.0%
	Total No. of Buildings under MBIS Year 2014 and % in terms of no. of buildings in 2014	45	10	0	0	55	Ranked No. 18 (2014)
		9.4%	2.1%	0.0%	0.0%	11.5%	
	Total No. of Buildings under MBIS by Year 2024 and % in terms of no. of buildings in 2014	113	25	31	10	179	Ranked No. 18 (2024)
		23.5%	5.2%	6.5%	2.1%	37.3%	
	Private Buildings under MBIS (Movement 2014–2024)	68	15	31	10	124	Ranking remains unchanged
		14.2%	3.1%	6.5%	2.1%	25.8%	

Note: For Private Buildings Non–Domestic and Domestic Buildings Over 3 storeys

MBIS, making the district the 18th among the 18 districts. In this aspect, by 2024, the projected number shall increase to 179 (37.3%), and the ranking shall be the last among the 18 districts (i.e., no. 18).

Comparatively speaking, Sai Kung is one of the youngest districts in the territory. On or before 1956, no private building was constructed. Up to 1971, there were only 9 private buildings in the district. In the period from 1987 to 2006, 380 private buildings (80.3%) were completed. In terms of size distribution (number of building units) of private buildings in the district, two characteristics are observed. First, the proportions of small-sized and medium-sized buildings were relatively low. As compared with the averages of 64.5% and 23.6% in the territory, the proportions of small-sized and medium- sized private buildings were 44.2% and 10.6% respectively. Second, the proportions of large-sized and extra-large-sized private buildings were much higher than that of the territory. As compared to the ones of 10.2% and 1.7% in the territory, the proportions of large-sized and extra-large-sized private buildings in the district were 34.4% and 10.8% respectively.

As of 31 December 2014, out of 55 private buildings (11.5%) falling under the MBIS, 45 were small-sized buildings (9.4%) and 10 were medium-sized buildings (2.1%). In the next 10 years, the projected number of aged private buildings falling under the MBIS will increase to 179 (37.3%). This represents a net increase of 124 (25.8%) in the next decade. The distribution of aged private buildings falling under the MBIS will be 113 for small-sized buildings (23.5%), 25 for medium-sized buildings (5.2%), 31 for large-sized buildings (6.5%) and 10 for extra-large-sized buildings (2.1%). The ranking, in terms of ageing rate (25.8%) of aged private buildings from 2014 to 2024, shall be the 8th.

Islands District, Mui Wo, Ngan Wan Road

5.2.18 Islands District

Key Features

(a) Ranked 18th in 2014 in number of private buildings with a total of 303 private buildings in the district

(b) Ranked 17th in 2014 with 92 (30.4%) buildings falling under the MBIS

(c) Shall rank 17th in 2024 with 242 buildings (79.9%) falling under the MBIS

(d) Changes in 10 years' time

- Ranking remains unchanged in terms of number of private buildings falling under the MBIS

- Projected number of buildings under the MBIS shall increase by 150

- Shall rank 2nd in terms of ageing rate (49.5%) of private buildings under the MBIS

Figure 5.18
Cumulative No. of Buildings (Per Unit Group)
in Islands District (As of 31 December 2014)

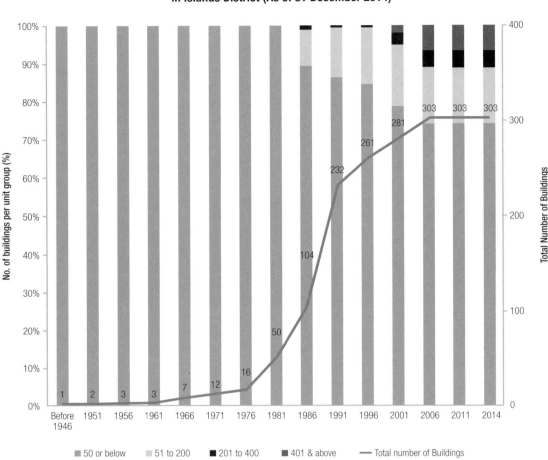

Note: For Non-Domestic Private Buildings and Domestic Buildings over 3 Storeys

Comment

The Islands District Council (District Council, 2014) highlights that "the Islands District consists of more than twenty islands of various sizes. The larger ones are Lantau Island, Lamma Island, Cheung Chau, Peng Chau and Po Toi. The Islands District, with 17,600 hectares of land area, is the largest among the 18 administrative districts in Hong Kong. Its population is around 151, 000 with the majority of which reside in the Tung Chung New Town, while the rest live in Cheung Chau, Peng Chau, Lamma Island, Tai O, Mui Wo, Tung Chung, South Lantau and Discovery Bay, etc."

The Islands District, with 303 private buildings, ranked 18th in 2014 in terms of number of private buildings among the 18 districts. As of 31 December 2014, there was a total of 92 private buildings (30.4%) in the district, aged 30 or above, falling under the

Table 5.18
Number of Private Buildings (Unit Distribution Per Year Group)
in Islands District (As of 31 December 2014)

		No. of Private Buildings (Per Unit Groups)					
		50 or above	51 or 200	201 to 400	401 and above		Percentage (%)
Year of Completion		Small	Medium	Large	Extra Large	Total	
1	2012–2014	0	0	0	0	0	0.0%
2	2007–2011	0	0	0	0	0	0.0%
3	2002–2006	3	0	5	14	22	7.3%
4	1997–2001	1	6	8	5	20	6.6%
5	1992–1996	20	9	0	0	29	9.6%
6	1987–1991	108	20	0	0	128	42.2%
7	1982–1986	43	10	1	0	54	17.8%
8	1977–1981	34	0	0	0	34	11.2%
9	1972–1976	4	0	0	0	4	1.3%
10	1967–1971	5	0	0	0	5	1.7%
11	1962–1966	4	0	0	0	4	1.3%
12	1957–1961	0	0	0	0	0	0.0%
13	1952–1956	1	0	0	0	1	0.3%
14	1947–1951	1	0	0	0	1	0.3%
15	≦1946	1	0	0	0	1	0.3%
Total No. of Buildings (Per Unit Group)		225	45	14	19	303	
% in Islands (Per Unit Groups)		74.3%	14.9%	4.6%	6.3%		100.0%
Total No. of Buildings under MBIS Year 2014 and % in terms of no. of buildings in 2014		85	6	1	0	92	Ranked No. 17 (2014)
		28.1%	2.0%	0.2%	0.0%	30.4%	
Total No. of Buildings under MBIS by Year 2024 and % in terms of no. of buildings in 2014		207	34	1	0	242	Ranked No. 17 (2024)
		68.3%	11.2%	0.3%	0.0%	79.9%	
Private Buildings under MBIS (Movement 2014–2024)		122	28	0	0	150	Ranking remains unchanged
		40.3%	9.2%	0.1%	0.0%	49.5%	

Note: For Private Buildings Non–Domestic and Domestic Buildings over 3 storeys

MBIS. The Islands also ranked last among the 18 districts. In this aspect, over the next decade, the projected number will increase to 242 (79.9%), and the ranking shall remain unchanged.

Comparatively speaking, the Islands is also one of the youngest districts in the territory. By 1976, there were only 16 private buildings in the district. In the period from 1977 to 1996, 245 (80.9%) private buildings were completed. In terms of size distribution (number of building units) of private buildings in the district, two characteristics are observed. First, the proportions of small-sized and extra-large-sized buildings were relatively high. There were 74.3% and 6.3% of small-sized and extra-large-sized private buildings in the district. As compared with the ones of 64.5% and 1.7% the territory, the proportions of small-sized and extra-large-sized private buildings in the district were 74.3% and 6.3% respectively. Second and conversely, the proportions of medium-sized and large-sized private buildings were lower. As compared with the ones of 23.6% and 10.2% in the territory, the proportions of medium-sized and large-sized private buildings were 14.9% and 4.6% respectively.

As of 31 December 2014, majority of building stocks falling under the MBIS were small-sized buildings. In the next 10 years, the projected number of aged private buildings falling under the MBIS will increase to 242 (79.9%). This represents a net increase of 150 in the next decade. The distribution of aged private buildings falling under the MBIS will be 207 for small-sized buildings (68.3%), 34 for medium-sized buildings (11.2%) and one for large-sized buildings (0.3%). The ranking, in terms of ageing rate (49.5%) of aged private buildings from 2014 to 2024, shall be the first.

5.3 Conclusions

Among the 18 districts in Hong Kong, Central and Western had 3,162 private buildings, the most among the districts. Yau Tsim Mong ranked second, with 3,041 private buildings. Sai Kung and the Islands were ranked penultimate and last, with only 480 and 303 private buildings respectively.

Yau Tsim Mong was the most densely aged district under the MBIS, with 2,478 private buildings as at 31 December 2014. Over the next decade, the projected number of private buildings falling under the MBIS in Yau Tsim Mong shall increase to 2,738. However, the Central and Western District—with 2,902 private buildings—will replace Yau Tsim Mong as the district with the most aged buildings.

Sai Kung was the "youngest" district in 2014 in terms of density of aged buildings falling under the MBIS. Out of its total number of 480 private buildings, only 55 have reached the age of 30 or above in 2014.

Finally, the rates of ageing in the Islands, Sha Tin and Tuen Mun in the next decade will become remarkably high. The density of the Islands, Sha Tin and Tuen Mun will increase from 30.4%, 25.9% and 32.4% to 49.5%, 41.1% and 51.4%, respectively.

References

District Council (2014). *District Highlights*. District Council of HKSAR. Available at http://www.districtcouncils.gov.hk/kt/english/info/highlight_01.html (Accessed on 10 December 2014)

CHAPTER 6

Analyses of Registered Inspectors and RI-eligible Building Professionals under MBIS

6.1 Introduction

This chapter provides a review on the requirements and administrative procedures of registering as an Registered Inspector (RI) under the MBIS. Detailed analyses on the supply pool of Authorised Persons (APs), Registered Structural Engineers (RSEs) as well as the number of potential building professionals in the relevant professional institutions and registration boards are conducted to portray the existing supply of professional workforce towards the MBIS. The building professionals are eligible to register themselves as RIs subject to their obtainment of prescribed qualifications and experience set forth under Section 3 of Cap 123A—Building (Administration) Regulations (B(A)R s.3).

   This chapter will further analyse the procedures, application, admission and failure rates of registering as RIs via the pathways of APs, RSEs and Registered Architects (RAs), Registered Professional Engineers (RPEs) and Registered Professional Surveyors (RPSs). The above figures can be used to estimate the potential supply of RIs available in the workforce market which will be discussed in Chapter 8.

6.2 Registered Inspectors (RIs)— Prescribed Qualifications and Experience

According to the Practice Note for Authorised Persons, Registered Structural Engineers and Registered Geotechnical Engineers (PNAP APP-7), the following professional bodies are considered having obtained "prescribed qualifications" for the purpose of registering themselves as Registered Inspectors in the Inspectors' Registry maintained by the Buildings Department under B(A)R s.3 (Buildings Department, 2011):

   (a)   Authorised Persons (AP(A), AP(E) and AP(S));

   (b)   Registered Structural Engineers (RSE);

   (c)   Registered Architects (RA);

   (d)   Registered Professional Engineers (RPE) in building, structural, civil, building services (building) and materials (buildings); and

   (e)   Registered Professional Surveyors (RPS) in building surveying or quantity surveying divisions.

   In addition to the prescribed professional qualifications, applicant of RI is also required to have the prescribed experience and/or pass a professional interview under PNAP APP-7 for registration as an RI as follows (Buildings Department, 2011):

- s.4(b)—For AP or RSE, within the seven years preceding the date of application, he must have experience gained in Hong Kong as an AP, RSE, RA, RPE or RPS in any building repair and maintenance project; or

- s.4(c)—For RA, RPE in the building or structural engineering discipline or RPS in the building surveying division, he must, for a period or periods in aggregate if not less than one year, within three years preceding the date of application, have practical experience in building construction, repair and maintenance gained in Hong Kong that the IRC considers appropriate; or

- s.4(d)—For RPE in the civil, building services (building) or materials (building) engineering discipline or RPS in the quantity surveying division, he must, for a period or periods in aggregate of not less than three years and of which at least one year falls within the three years preceding the date of application, have practical experience in building construction, repair and maintenance gained in Hong Kong that the Inspectors Registration Committee (IRC) considers appropriate.

6.3 Recognised Building Professional

6.3.1 Corporate Members of the HKIA/HKIE/HKIS and Registered Members of the ARB/ERB/SRB

According to the membership statistics published on the websites of the Hong Kong Institute of Architects (HKIA), the Hong Kong Institution of Engineers (HKIE) and the Hong Kong Institute of Surveyors (HKIS), the analysis on such database can portray a real picture on the number of professionals who are eligible to register themselves as RI. As of 22 February 2015, there were a total of 17,600 corporate members in relevant disciplines or divisions of the HKIA, HKIE and HKIS whom may ultimately be eligible to register themselves as RIs in the Inspectors' Registry of the Buildings Department (HKIA, 2015; HKIE, 2015; HKIS, 2015). Their registration is subjected to the relevant registration boards and obtainment of prescribed experience under B(A)R s.3. The numbers of building professionals in the HKIA and respective disciplines and divisions of the HKIE and HKIS are provided in Table 6.1.

As of 22 February 2015, there were 10,398 corporate members in the HKIE, 3,944 in the HKIS and 3,258 in the HKIA. Therefore, the total number of potential RI-eligible building professionals is 17,600.

Further study on membership statistics published on the websites of the Architects Registration Board (ARB, 2015), the Engineers Registration Board (ERB, 2015; HKIE, 2015) and the Surveyors Registration Board (SRB, 2015) revealed that out of the total 17,600 corporate members, only 9,385 had registered in relevant registration boards recognised by the Buildings Department.

Table 6.1
Statistics of Potential RI-Eligible Building Professionals
(HKIA, HKIE and HKIS) (As of 22 February 2015)

	Professional Institutions	Architects	Engineers Disciplines	Surveyors Divisions	Total
1	Hong Kong Institute of Architects (HKIA)	3,258	0	0	3,258
2	Hong Kong Institute of Engineers (HKIE)				10,398
	a Building Discipline	0	395	0	
	b Structural Discipline	0	2,682	0	
	c Civil Discipline	0	5,723	0	
	d Building Services Discipline	0	1,516	0	
	e Materials Discipline	0	82	0	0
3	Hong Kong Institute of Surveyors (HKIS)			3,944	
	a Building Surveying Division	0	0	1152	
	b Quantity Surveying Division	0	0	2,792	
	Total:	3,258	10,398	3,944	17,600

Source: Hong Kong Institute of Architects (HKIA), 2015; Hong Kong Institution of Engineers (HKIE), 2015; Hong Kong Institute of Surveyors (HKIS), 2015

Regarding the percentage of corporate members, 3,202 of the 3,258 professional architects (98%) are registered on the ARB. Out of the total of 3,944 professional surveyors, 2,089 (53%) had registered themselves on SRB. However, only 4,786 out of the 10,398 professional engineers (46%) on ERB. The overall average registration rate on relevant registration boards was 57% (10,077 out of the total 17,600 members). However, according to s.4 of PNAP APP-7 (Buildings Department, 2011), RPE members of Building Services and Materials Disciplines are not directly eligible to register themselves as RIs unless they are included on the relevant Sub-lists of RPE, namely Building Services (Building) and Materials (Building). It is noticed that the rate of inclusion in these two RPE Sub-lists is still very low. Out of the total of 790 members of RPE Building Services Discipline and 25 members of RPE Materials discipline, only 112 members (14%) and 11 members (44%) have included themselves on the Sub-lists of Building Services (Building) and Materials (Building).

Taking into account the restriction of RPE Sub-list for Building Services (112) and Materials Disciplines (11) under s.4 of PNAP APP-7 (Buildings Department, 2011), the total RI-eligible building professionals then dropped from 10,077 to 9,385. Detailed

Table 6.2
Statistics of RI-Eligible Building Professionals (ARB, ERB and SRB)
(As of 22 February 2015)

	Registration Boards	RA	RPE	RPE (Sub- list)	RPS	Total (RA/ RPS/RPE)	Total (RA/ RPS/RPE/ SRB) (Building)
1	Architects Registration Board (ARB)	3,202	0	0	0	3,202	3,202
2	Engineers Registration Boards (ERB)					4,786	4,094
	a Building Discipline	0	213	N/A	0		
	b Structural Discipline	0	1,671	N/A	0		
	c Civil Discipline	0	2,087	N/A	0		
	d Building Services Discipline	0	790	112	0		
	e Materials Discipline	0	25	11	0		
3	Surveyors Registration Board (SRB)					2,089	2,089
	a Building Surveying Division	0	0		840		
	b Quantity Surveying Division	0	0		1,249		
	Total (A):	3,202	4,286	123	2,089	10,077	9,385
4	Number of Building Professional in HKIA, HKIE and HKIS (B)	3,258	10,398	N/A	3,944	17,600	17,600
	Percentage of Registration (%) = (A)/ (B)	98%	46%	N/A	53%	57%	53%

Source: Architects Registration Board (ARB), 2015; Engineers Registration Board (ERB), 2015; Surveyors Registration Board (SRB), 2015

numbers of the RI-eligible building professionals in respective registration boards (ARB, ERB and SRB) are listed in Table 6.2.

Nevertheless, the actual number of RI-eligible building professionals dropped even further because of the multiple ownerships of numerous RI-eligible building professionals in several recognised registration boards. Analysis reveals that as of 22 February 2015, the net number of RI-eligible building professionals in the relevant divisions, disciplines and Sub-lists of the ARB, ERB and SRB was only 8,650.

In fact, the number matches the assessment made by the Government, which claims that about 8,000 building professionals are qualified to register as RI, including RAs,

Figure 6.1
Number of RI-eligible Building Professionals and Multiple Memberships
in Different Recognised Registration Boards (As of 22 February 2015)

Number of Registered Inspector (RI) Eligible Building
Professionals in ARB, ERB and SRB

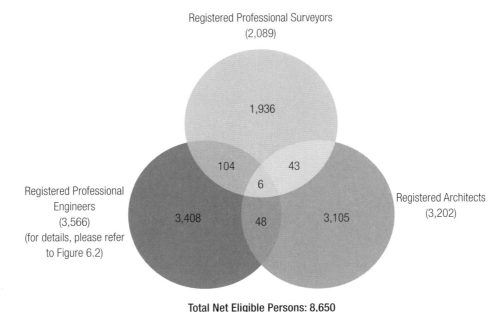

Total Net Eligible Persons: 8,650

Note: Figures in brackets denote the total number of RI eligible professionals (i.e., RA/RPE/RPS) in relevant registration
   boards
Source: Architects Registration Board (ARB) 2015; Engineers Registration Board (ERB) 2015; Surveyors Registration
   Board (SRB) 2015

RPEs and RPSs in relevant disciplines/divisions (Legislative Council, 2011d, 2011e; Buildings Department, 2012a, 2012b). Figures 6.1 and 6.2 illustrate the numbers of registered building professionals and their multiple memberships across the ARB, RPE, RPS and different disciplines of the RPE.

Figure 6.1 shows that in 2015 there were 104 RI-eligible building professionals holding double memberships as RPE and RPS, 43 as RPS and RA and 48 as RPE and RA respectively. Six members held triple memberships as RA, RPE and RPS.

The above analysis reveals that the number of building professionals of RA, RPE and RPS would drop further from 9,385 registered professional members to 8,650 (7.83% downward) after applying the effects of multiple professional registrations. Furthermore, multiple memberships of RPE among different disciplines are shown in Figure 6.2. There

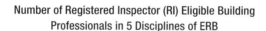

Figure 6.2
Number of RI-eligible Building Professionals and Multiple Memberships
in Different Disciplines of ERB (As of 22 February 2015)

Number of Registered Inspector (RI) Eligible Building
Professionals in 5 Disciplines of ERB

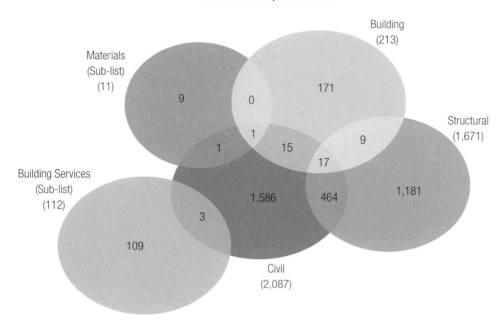

Total Net Eligible Persons: 3,566

Note: Figures in brackets denote the total number of RI eligible professionals in different recognised Disciplines of ERB
Source: Engineers Registration Board (ERB)

were as many as 464 members holding double membership in both Structural and Civil Disciplines in the year of 2015. There were 17 members holding triple memberships in Building, Structural and Civil Disciplines.

6.3.2 Authorised Persons and Registered Structural Engineers

As mentioned above, s.4(a) of PNAP APP-7 stipulates that "a person must not be included in the Inspectors' Registry unless he is an AP, RSE, RA, RPE …". and s.4(b) further elicits that "for AP or RSE, within the 7 years preceding the date of application, he must have experience gained in Hong Kong as an AP, RSE, RA, RPE or RPS …" (Buildings Department, 2011).

Table 6.3
Authorised Persons' Register and Structural Engineers' Register
in the Buildings Department (As of 22 February 2015)

	Buildings Department's Register	Sub- Total	Total
1	Authorised Persons (APs)		1,518
a.	List of Architects—AP(A)	1,182	
b.	List of Engineers—AP(E)	142	
c.	List of Surveyors—AP(S)	194	
2	Registered Structural Engineers (RSE)		423
	Total:		1,941

Source: Buildings Department, 2015a

As of 22 February 2015, there were 1,941 listed members in the Authorised Persons' Register and Structural Engineers' Register (Buildings Department, 2015a). Breakdowns of the listed members are provided in Table 6.3.

It is reckoned that, as of 22 February 2015, out of the total 1,941 APs and RSEs listed in the Registers of the Buildings Department under s.3(1), s.3(2) and s.3(3) of the Building (Administration) Regulations, the majority are listed in two or more categories. As shown in Figure 6.3, there were as many as 133 members holding double registration in both AP(E) and RSE. Two held double registration as AP(A) and RSE and triple registration as AP(E), AP(S) and RSE. After netting off multiple memberships as AP(A), AP(E), AP(S) and RSE, the net number was 1,795 only. This figure again dovetails with the Government's estimate of 1,800 APs and RSEs in June 2011 (Legislative Council, 2011c).

It is of paramount importance to note that the number of RI-eligible building professionals will not increase with the additional 1,795 APs and RSEs listed in the Buildings Department. The reason is that under s.3(a) and s.3(b) of PNAP APP-7 (Buildings Department, 2011), the qualifications prescribed under B(A)R s.3 of the BO for registration as AP requires the applicant to be a registered architect (RA), a registered professional engineer (RPE) in the structural or civil engineering discipline or a registered professional surveyor (RPS). Whereas to register as a RSE, a person has to be an RPE in the structural or civil engineering discipline. In this connection, the numbers of APs and RSEs are virtually absorbed into the total pools of RAs, RPEs and RPSs respectively.

Figure 6.3
Number of RI-eligible Building Professionals (APs and RSEs) and Multiple Memberships Listed in the Buildings Department (As of 22 February 2015)

Number of Registered Inspector (RI) Eligible Building Professionals—
APs and RSE in Buildings Department

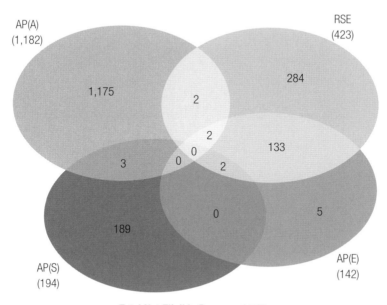

Total Net Eligible Persons: 1,795

Source: Buildings Department, 2015b

6.4 Status of Registrations of Registered Inspectors

6.4.1 Registered Inspectors' Registration Rates

Given the enthusiastic participation and responses from members of the HKIA, HKIE and HKIS in the course during discussion of the implementation of the MBIS, the three professional institutions indicated that there should be adequate building professionals for registration as RIs (Legislative Council, 2011a, 2011e).

The Hong Kong Government also assessed that there would be around 6,500 RI-eligible building professionals by expanding the pool to include RAs, RPEs of relevant disciplines and RPSs of relevant divisions. If 15% to 20% in such pools (i.e., about 950 to 1,300) were registered, there would be a sufficient amount of RIs meeting the workforce demand of the MBIS (Legislative Council, 2011a, 2011c, 2012a).

Table 6.4
Registered Inspectors (RIs) Registrations (As at 22 February 2015)

Profession	No. of RIs Registered in relevant Inspectors' Registers (i.e., RI(A), RI(E) and RI(S))	No. of Members in relevant Professional Institutions (i.e., HKIA, HKIE and HKIS)		No. of Members in relevant Registration Boards (i.e.,ARB, ERB and SRB)		No. of authorised persons and Registered Structural Engineers (i.e. AP(A), AP(E), AP(S) and RSE)
		No.	RI Registration Rate	No.	RI Registration Rate	
	(a)	(b)	(c)=(a)/(b) x 100%	(d)	(e)=(a)/(d) x 100%	(f)
Architects	136	3,258	4.7%	3,202	4.3%	1,182
	(32%)	(19%)		(36%)		(60%)
Engineers	157	10,398	1.5%	3,566 [#2]	4.40%	565 [#1]
	(38%)	(59%)		(40%)		(30%)
Surveyors	127	3,944	3.2%	2,089	6.1%	194
	(30%)	(22%)		(24%)		10%
Total	420	17,600	2.4%	8,857	4.7%	1,941
						(100%)

Note [#1]:   refers to the total numbers of AP(E) and RSE

Note [#2]:   The eligible professional engineer across all discipline is 4,094. Refer to Figure 6.2 , the net eligible number is only 3,566 (after the deduction of multiple membership)

Remarks: For the sake of evaluating registration rates, all multiple memberships in Professional Institutions (HKIA, HKIE, HKIS), Registration Boards (ARB, ERB, SRB) and Authorised Persons (AP(A), AP(E), AP(S)) and Registered Structural Engineers (RSE) have not been adjusted

Source: Buildings Department, 2015a, 2015b; Hong Kong Institute of Architects (HKIA), 2015; Hong Kong Institution of Engineers (HKIE), 2015; Hong Kong Institute of Surveyors (HKIS), 2015

According to the Buildings Department (2015b), there were only 420 RIs whom listed in the Inspectors' Registry of the Buildings Department as of 22 February 2015. Breakdowns of the RI registration in Lists of Architects, Engineers and Surveyors and registration rates in respect of different professions are provided in Table 6.4.

The proportion of registration among the three RI-eligible building professions, namely architects, engineers and surveyors is roughly equal. As at 22 February 2015, out of the total 420 RIs registered, 136 RIs (33%) originated from the group of professional architects; 157 (38%) from professional engineers and 127 (30%) from professional surveyors.

Figure 6.4
Registered Inspectors (RIs) Registration (As of 22 February 2015)

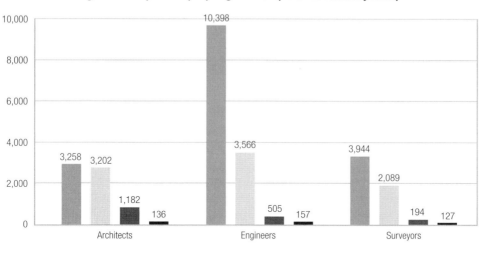

■ No. of Professional Members in relevant Professional Institutions (i.e., HKIA, HKIE and HKIS)
▨ No. of Professional Members in relevant Registration Boards (i.e., ARB, ERB and SRB)
■ No. of Authorised Persons and Registered Structural Engineers (i.e., AP(A), AP(E), AP(S) and RSE)
■ No. of RIs registered in the Inspectors' Registry of the Buildings Department

Source: Buildings Department, 2015a, 2015b; Hong Kong Institute of Architects (HKIA), 2015; Hong Kong Institution of Engineers (HKIE), 2015; Hong Kong Institute of Surveyors (HKIS), 2015

Although professional engineers make up for a large proportion of (157 in number) RI-eligible building professionals, their registration rate was the lowest compared with the total number of RI-eligible RPEs. As shown in Table 6.4 and Figure 6.4, the most active group of building professionals registered as RI is surveyor. Out of 2,089 Registered RPSs on the Surveyors Registration Board (SRB), 127 RPSs listed themselves on the Registered Inspectors (RIs) (List of Surveyors), i.e., approximately 6.1% of the total RI-eligible professional surveyors. The next most active group is architect, with 136 Registered Architects (RA) listed themselves as RI(A), which is approximately 4.3% of the total RI-eligible professional architects. The least active group was engineer. Out of the 3,566 net RI-eligible RPE, only 157 RPEs listed themselves as RI(E), which is approximately 4.4% of the total eligible professional engineers.

The above RIs registration rates represent the "overall result" of RI registration (420 out of 8,857, 4.7%) in respect of the total number of members in relevant Disciplines/Divisions of Professional Institutions (i.e., HKIA, HKIE and HKIS) and Registration Boards (i.e., ARB, ERB and SRB). For the sake of evaluating the actual RI registration rates, more detailed studies will be discussed in a later Chapter to analyse the actual registration rates after netting off the effect of multiple memberships in different disciplines/divisions of the Registration Boards and APs and RSEs.

6.4.2   Qualifications Prescribed for Registration as Registered Inspectors

Under Section 3 of B(A)R for registration as an RI, the qualifications prescribed for (1) AP and RSE, (2) RA, RPE in the Building or Structural Engineering Disciplines or RPS in the Building Surveying Division and (3) RPE in the Civil, Building Services (Building) or Materials (Building) Engineering Discipline or RPS in the Quantity Surveying Division are not the same.

Qualifications for the purpose of registering as RIs are detailed in PNAP APP-7 (Buildings Department, 2011, p.2). The differences in requirements between AP and RSE, and that of RA, RPE and RPS are compared and summarised in Table 6.5.

As shown in Table 6.5, the key difference between the groups of AP and RSE, and RA, RPE and RPS is that no Inspectors Registration Committee's (IRC) vetting requirement is set for AP's and RSE's practical experience. This direct entry route for APs and RSEs is stipulated in s.3(7AA)(a) of the Buildings Ordinance (BO). For the purpose of registering

Table 6.5
Differences of Prescribed Qualifications
between AP/RSE and RA/RPE/RPS

Group of Professionals	Required Length of Experience	Practical Experience Vetted by IRC
1. AP and RSE (PNAP APP-7 s.4(b))	Within the 7 years preceding the date of application, he must have experience gained in Hong Kong as an AP, RSE, RA, RPE or RPS in any building repair and maintenance project	Not required
2. RA, RPE in the Building or structural engineering descipline or RPS in the building surveying division (PNAP APP-7 s.4(c))	He must, for a period or periods in aggregate of not less than 1 years within 3 years preceding the date of application.	have practical experience in building construction, repair and maintenance gained in Hong Kong that the IRC considers appropriate. (i.e., examination of qualifications and experience and professional interview by the IRC)
3. RPE in the civil, building services (building) or materials (building) engineering discipline or RPS in the quantity surveying division (PNAP APP-7 s.4 (d))	He must, for a period or periods in aggregate of not less than 3 years and of which at least 1 year falls within the 3 years preceding the date of application	have practical experience in building construction, repair and maintenance gained in Hong Kong that the IRC considers appropriate. (i.e., examination of qualifications and experience and professional interview by the IRC)

Source: Buildings Department, 2011

as a RI, an AP or a RSE only requires one to show his/her experience in Hong Kong as an AP, RSE, RA, RPE or RPS in any building repair and maintenance project within the seven years preceding the date of the application. Conversely, with the exception of building professionals under s.3(7AA)(b) of the BO, experience of all RI-eligible RAs, RPEs or RPSs, regardless of their disciplines or divisions, have to be vetted by the IRC. According to PNAP APP-7 (Buildings Department, 2011), the IRC will, in addition to examining the qualifications and experience, conduct professional interviews for all applications from the groups of RA, RPE and RPS.

It is considered that the vetting process conducted by the IRC, the examination of qualifications, experience and professional interviews, for all applications from the groups of RA, RPE and RPS may be one of the key factors for the limited number of RI applications and registrations.

6.4.3 Analyses of Registered Inspectors' Applications and Admission Rates

To find out the admission rate of building professionals being registered as RI, the registration figures was collected from local building authority—the Buildings Department. For instance, as of 30 November 2012, the total number of applications for Registered Inspectors received was 439, among which three applicants withdrew (Buildings Department, 2012d). Taking out 51 applications which were being processed at the material time, the net number of application for RI registration was 385.

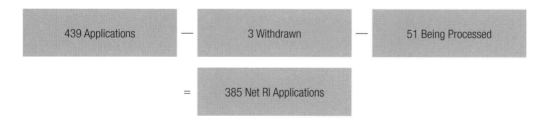

Out of these net RI applications, 29 were admitted under "Transitional Arrangement", a special arrangement under s.3 of B(A)R (Buildings Department, 2012d). Under Transitional Arrangement, a person may be included on the list of Inspectors' Registry of the Buildings Department without having the recommendation by an Inspectors Registration Committee (IRC) if the person is a RA, RPE or RPS whom possesses not less than five years of experience in building design, construction, repair and maintenance before the nomination. Such Transitional Arrangement is to allow a small number of experienced professionals to become RIs as soon as the Amendment Ordinance comes into effect so the Inspectors Registration Committee could be found at the outset to

Table 6.6
Registered Inspectors Admitted under Transitional Arrangement

Registered Inspectors	AP / RSE	RA / RPE / RPS	Total
List of Architects	9	1	10
List of Engineers	7	2	9
List of Surveyors	9	1	10
Total:	25	4	29

scrutinise the RI applications (Development Bureau, 2011b). Details of the RIs admitted under the Transitional Arrangement are shown in Table 6.6.

Taking out the 29 RIs admitted under Transitional Arrangement from the 385 net applications, the net RI applications was 356. Likewise, removing the 29 RIs admitted under Transitional Arrangement from the 308 approved RIs, the net approved RIs as of 30 November 2012 was 279. Therefore, the overall admission rate based on the number of approved RIs and the net number of applications (i.e., excluding applicants under Transitional Arrangement) was

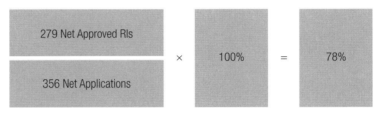

Overall Admission Rate

Nevertheless, Buildings Department further advised that the total number of applications rejected or deferred was 77. Based on this figure, the failure rate can also be derived as follows:

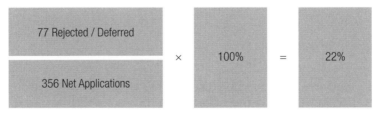

Overall Failure Rate

However, further studies reveal that the above overall admission rate cannot truly reflect the inherent problems of RI registrations. Table 6.7 below tabulates the statistics on the progress of RI registrations in terms of (i) Groups of RI-eligible professional bodies (i.e. Architects, Engineers and Surveyors) and (ii) AP and RSE, and RA, RPE and RPS.

Table 6.7

Status of RI Applications and Registrations (As of 30 November 2012)

RI Application and Approval Status		RI (Architects)			RI (Engineers)			RI (surveyors)			Total RIs		
		AP/RSE	RA	Sub-total	AP/RSE	RPE	Sub-total	AP/RSE	RPS	Sub-total	AP/RSE	RA/RPE/RPS	Sub-total
No. of Applications	(a)	117	9	126	102	90	192	65	56	121	284	155	439
No. of Withdrawals	(b)	1	1	2	1	0	1	0	0	0	2	1	3
No. in Progress	(c)	13	1	14	11	19	30	1	6	7	25	26	51
Sub-Total	(d) = (a)-(b)-(c)	103	7	110	90	71	161	64	50	114	257	128	385
No. Admitted under Transitional Arrangement	(e)	9	1	10	7	2	9	9	1	10	25	4	29
Net Applications	(f) = (d)-(e)	94	6	100	83	69	152	55	49	104	232	124	356
No. of Approvals	(g)	103	3	106	89	20	109	63	30	93	255	53	308
No. Admitted under Transitional Arrangement	(h)	9	1	10	7	2	9	9	1	10	25	4	29
Net Approved	(i) = (g)-(h)	94	2	96	82	18	100	54	29	83	230	49	279
Passing Rate	(j) = (i)/(f) x 100%	100%	33%	96%	99%	26%	66%	98%	59%	80%	99%	40%	78%
No. Rejected or Deferred etc.	(k)	0	4	4	1	51	52	1	20	21	2	75	77
Failure Rate	(l) = (k)/(f) x 100%	0%	67%	4%	1%	74%	34%	2%	41%	20%	1%	60%	22%

Table 6.8
Eligible Professional Bodies for Registered Inspectors

Registration Boards	Registered Professionals
1. Architects Registration Board (ARB)	Registered Architects (RA)
2. Engineers Registration Board (ERB)	Registered Professional Engineers (RPE)
	a. Building Discipline
	b. Structural Discipline
	c. Civil Discipline
	d. Building Services (building) Discipline
	e. Materials (Building) Discipline
3. Surveyors Registration Board (SRB)	Registered Professional Surveyors (RPS)
	a. Building Surveying Division
	b. Quantity Surveying Division

Source: Buildings Department, 2011

6.4.4 RI Analyses Based on Groups of RI-eligible Building Professionals

Under s.4 of PNAP APP-7 (Buildings Department, 2011), registered members in the 8 disciplines and divisions from the Architects Registration Board (ARB), Engineers Registration Board (ERB) and Surveyors Registration Board (SRB) are eligible to apply for registration of Registered Inspectors subject to fulfillment of prescribed qualifications set forth under s.3 of B(A)R of the BO.

As shown in Table 6.8, the eligible professional bodies for registration as an RI shall be architects, engineers and surveyors. Mindful observation also reveals that the greatest number of RI applications as of 30 November 2012 was flowing from the group of engineers (192). Architects (126) ranked second and surveyors (121) was the last. After netting off the RI applications and admissions under Transitional Arrangement per s.3 of B(A)R and the numbers of withdrawals and applications in progress, the net number of applications for the purpose of analysing the admission rates from the groups of architects, engineers and surveyors are 100, 152 and 104 respectively.

6.4.5 RIs Admission Rates Analyses Based on Professions

Below are the analyses of RI applications and the admission rates of the three professional bodies.

(a) Architects—As shown in Table 6.7, 96 out of 100 net applications for RI(A) under the group of Architects (excluding RIs admitted under Transitional Arrangement) were admitted. The admission rate is therefore 96%.

(b) Engineers—100 out of the total 152 net applications for RI(E) under the group of Engineers (excluding RIs admitted under Transitional Arrangement) were admitted. The admission rate is therefore 66%.

(c) Surveyors—83 out of the total 104 net applications for RI(S) under the group of Surveyors (excluding RIs admitted under Transitional Arrangement) were admitted. The admission rate is therefore 80%.

(d) Overall Result—The overall admission rate for the 3 groups of professional bodies (architects, engineers and surveyors) is 78%.

6.4.6 RIs Admissions Rate based on Groups of AP and RSE and RA, RPE and RPS

Further analyses of RI application admission rates based on groups of AP and RSE, and RA, RPE and RPS reveal that the results are astonishing and vary greatly among different groups of building professions. The analyses further divide the groups into AP(A) and RA, AP(E) and RSE, RPE, and AP(S) and RPS.

(a) Architects (AP(A) and RA)

- Authorised Persons (List of Architects) (AP(A))—Out of the total 94 net applications for RI(A) under the group of AP(A) (excluding RIs admitted under Transitional Arrangement), 94 AP(A) were admitted. This means that the admission rate for AP(A) under the Authorised Persons (List of Architects) is 100%.

- Registered Architects (RA)—There were a total of six applications for RI(A) from RA and only two were admitted. The admission rate is therefore 33%.

(b) Engineers

- Authorised Persons (List of Engineers) (AP(E)) and Registered Structural Engineers (RSE)—82 out of 83 net applications for RI(E) under the group of AP(E) and RSE (excluding RIs admitted under Transitional Arrangement) were admitted. This means that the admission rate for AP(E) and RSE is 99%.

- Registered Professional Engineers (RPE)—There were a total of 69 applications for RI(E) from different disciplines of RPE and only 18 were admitted. The admission rate is therefore 26%.

(c) Surveyors

- Authorised Persons (List of Surveyors) (AP(S))—Out of the total 55 net applications for RI(S) under the group of AP(S) (excluding RIs admitted under Transitional Arrangement), 54 AP(S) were admitted. This means that the admission rate for AP and RSE under the List of Surveyors is 98%.

- Registered Professional Surveyors (RPS)—There were a total of 49 applications for RI(S) from the Divisions of Building Surveying and Quantity Surveying of RPS, 29 were admitted. The admission rate is therefore 59%.

(d) Overall Result

- Authorised Persons AP(A), AP(E), AP(S) and RSE—As shown in Table 6.6, out of the total 232 net applications for RIs under the group of AP and RSE (excluding RIs admitted under Transitional Arrangement), 230 AP and RSE were admitted. This means that the admission rate for AP and RSE under the Lists of Architects, Engineers and Surveyors is 99%.

- RAs, RPEs and RPSs—Out of the total 124 applications for RIs under the groups of RA, RPE and RPS, 49 candidates were admitted. The admission rate is therefore 40%.

6.4.7   Evaluation on RI's Admission Rate

Based on the above analyses, whilst the overall RI application admission rate of 78% (279 applications admitted out of 356 applications as of 30 November 2012) is fairly good, analyses reveal that the actual RI application and admission rates of some professional groups are fairly low.

For AP and RSE, 230 out of 232 applications were admitted. The result means that almost all applicants for RIs from the group of AP and RSE were admitted as RIs. The admission rates were:

AP(A)—100% (94 admissions out of 94 applications)

AP(E) and RSE—99% (82 admissions out of 83 applications)

AP(S)—98% (54 admissions out of 55 applications)

Overall—99% (230 admissions of 232 applications)

However, the admission rates for RI applications from the groups of RA, RPE and RPS is relatively unsatisfactory. The admission rates are as follows:

RA—33% (2 admissions out of 6 applications)

RPE—26% (18 admissions out of 69 applications)

RPS—59% (29 admissions out of 49 applications)

Overall—40% (49 admissions out of 124 applications)

Based on the above results and excluding the admissions under AP and RSE, the highest RI admission rate in terms of RI-eligible building profession is Surveyors with 59%. The second highest profession is Architects with 33%. The lowest admission rate in terms of profession is Engineers with only 26%.

In this connection, further analyses are made to investigate the reasons of the low admission rates in different professions, in particular with engineers and architects.

6.5 Conclusions

As of 22 February 2015, there were a total of 17,600 building professionals in relevant disciplines and divisions of the three professional institutions (the HKIA, HKIE and HKIS), among which 9,385 have registered themselves as RAs, RPEs and RPSs in relevant disciplines and divisions of the registration boards (the ARB, ERB and SRB). However, after netting off the effects of multiple memberships, the net number of RI-eligible building professionals was only 8,650.

The Government assessed that there would be around 6,500 RI-eligible building professionals and that if 15% to 20% of the 6,500 (i.e., about 950 to 1,300) were registered, there would be sufficient RIs to meet the demand. Reviewing the registration figures provided by local building authority - Buildings department, it revealed that the overall RIs admission rate was fairly good and the overall admission rate for APs and RSEs was up to 99% while the admission rates for individual professions were unsatisfactory, especially for RAs and RPEs whom are required to attend the professional interviews conducted by the Inspectors Registration Committee (IRC). The admission rates for these professions areas low as 33% (RA) and 26% (RPE).

References

Architect Registration Board (ARB) (2015). *The Register (Registered Architect List)*. Architect Registration Board. Available at http://218.188.25.84/ARB/English/TheRegister-A.php (Accessed on 22 February 2015)

Buildings Department (2011). *Registration of MBIS Registered Inspectors begins*, APP-7. Buildings Department of HKSAR. Available at http://www.bd.gov.hk/english/documents/news/20111229Bae.htm (Accessed on 21 March 2012).

Buildings Department (2012a). *Mandatory Building Inspection Scheme (Pamphlet on MBIS)*. Hong Kong: Buildings Department of HKSAR.

Buildings Department (2012b). *Mandatory Window Inspection Scheme (Pamphlet on MWIS)*. Hong Kong: Buildings Department of HKSAR.

Buildings Department (2012d). *Planning and Lands Branch—Task Force on Building Safety and Preventive Maintenance* (LM 1 to DEVB(PL-B)68/09/07) (e-message to author, 6 December 2012). Hong Kong: Buildings Department of HKSAR.

Buildings Department (2015a). *Authorized Person's Registers*. Buildings Department of HKSAR. Available at http://www.bd.gov.hk/english/inform/index_ap.html (Accessed on 22 February 2015)

Buildings Department (2015b). *Inspectors' Register*. Buildings Department of HKSAR. Available at http://www.bd.gov.hk/english/inform/index_ap.html (Accessed on 22 February 2015)

Development Bureau (2011b). *Examination of Estimates of Expenditure 2011–12* (Reply Serial No. DEVB(PL)(106), Question Serial No. 1194, 16 March 2011).Hong Kong: Development Bureau of HKSAR.

Engineers Registration Board (ERB) (2015). *Search Professional Engineer (R.P.E.) Enquiry*. Engineers Registration Board. Available at http://www.erb.org.hk/search.htm (Accessed on 22 February 2015)

Hong Kong Institute of Architects (HKIA) (2015). *Looking for Architects*. Hong Kong Institute of Architects. Available at http://www.hkia.net/en/LookingForArchitects/LookingForArchitects_01.htm (Accessed on 22 February 2015)

Hong Kong Institution of Engineers (HKIE) (2015). *HKIE Members and Mandatory Building Inspection Scheme* (e-mail message to author, 22 February 2015). Hong Kong: HKIE.

Hong Kong Institute of Surveyors (HKIS) (2015). *Find a Member*. The Hong Kong Institute of Surveyors. Available at http://www.hkis.org.hk/en/membership_find.php (Accessed on 22 February 2015)

Legislative Council (2011a). *Paper for the Bills on Buildings (Amendment) Bill 2010, Proposed amendments to include the new building safety initiatives* (LC Paper No. LS62/10-11, 17 May 2011). Hong Kong: Legislative Council of HKSAR.

Legislative Council (2011c). *Paper on Development Subcommittee on Building Safety and Related Issues, Meeting on 26 August 2011—Updated background brief on building safety* (LC Paper No. CB(1)2930/10-11(04), 26 August 2011). Hong Kong: Legislative Council of HKSAR.

Legislative Council (2011d). Legislative Council Brief—*Subsidiary Legislation for Implementation of Mandatory Building Inspection and Mandatory Window Inspection Scheme* (8 October 2011). Hong Kong: Legislative Council of HKSAR.

Legislative Council (2011e). Panel on Development Subcommittee on *Updated Background Brief on Mandatory Building Inspection Scheme and Mandatory Window Inspection Scheme* (LC Paper No. CB(1)137/11-12(06), 24 October 2011). Hong Kong: Legislative Council of HKSAR.

Legislative Council (2012a). Mandatory building and window schemes. *Official Record of Proceedings* (pp. 15016–15023, 13 June 2010). Hong Kong: Legislative Council of HKSAR.

Surveyors Registration Board (SRB) (2015). *Registered Professional Surveyors.*Surveyors Registration Board. Available at http://www.srb.org.hk/division.html (Accessed on 22 February 2015)

CHAPTER 7

Analyses of Registered Inspector Application, Registration and Failure Rates

7.1 Introduction

This chapter provides a critical evaluation on the numbers of applications and registrations of RIs. It includes analyses of admissions of RIs under the Transitional Arrangement, admissions under the pathway of AP and RSE in which professional interviews and vetting by the Inspectors Registration Committee (IRC) are not required and admissions under the pathway of RA, RPE or RPS in which professional interviews and vetting by IRC are required. With these analyses, factors leading to the low registration rates of RIs were identified. An in-depth study was conducted to analyse the registration rates in each discipline and division of the RI-eligible registration boards (the ARB, ERB and SRB), as well as their relationships with the scope of professional interviews.

7.2 Background for the Study on Low Registration Rates of Registered Inspectors

Based on the registration figures at 30 November 2012 provided by Buildings Department (2012d), a total of 308 RIs were listed in the Inspectors' Registry System. In Chapter 6, it also reveals that the overall RI registration rate of 78% (279 out of 356 RI's applications) is considered fairly well. The RI admission rates for the groups of AP and RSE are as high as 99% (230 out of 232 applications). On the contrary, the RI admission rates for the groups of RA, RPE and RPS appear to be very low.

In this regard, the following analyses are made to study the paramount reasons of low registration rates of RIs:

(a) Registered Inspectors (RIs) admission pathways;

(b) Analyses of the professional qualifications of RIs and scope of the professional interviews under PNAP APP-7.

7.3 Admission Pathways of Registered Inspectors (RIs)

Under section 3 of the Buildings Ordinance (BO), there are three admission pathways for a building professional to register to be a RI in the Inspectors' Registry of the Buildings Department, namely:

(a) RI admission under the Transitional Arrangement (s.3 (7AA)(a) of the BO);

(b) RIs admission not required to attend professional interviews (i.e., for RA, RPE or RPS holding the required additional qualifications of AP(A), AP(E), AP(S) and/or RSE)(s.3 (7AA)(b) of the BO); or

Table 7.1
Registered Inspectors Admission Pathways
and Number of Registered Inspectors (RIs) (As of 30 November 2012)

Registered Inspectors (RIs) Admission Pathways	RI(A)	RI(E)	RI(S)	Total	% of total number of RIs
1a. RIs admission under Transitional Arrangement with AP/RSE qualification	9	7	9	25	8%
1b. RIs admission under Transitional Arrangement without AP/RSE qualifications	1	2	1	4	1%
2. RIs admission under AP/RSE (i.e., not required to attend Professional Interviews)	94	82	54	230	75%
3. RIs admission under RA/PPE/RPS (i.e., required to attend Professional Interviews)	2	18	29	49	16%
Total :	106	109	93	308	100%

Source: Buildings Department, 2012c

(c)   RIs admission required to attend professional interviews (i.e., for RA, RPE or RPS without holding the required additional qualifications of AP(A), AP(E), AP(S) and/or RSE) (s.3 (7) of the BO).

To retrieve the current registered rates of building professionals as RIs, Table 7.1 outlines the number of admitted RI(A), RI(E) and RI(S) and the three different admission pathways for registering as Registered Inspectors (RIs) in the Inspectors' Registry of the Buildings Department.

7.3.1   Registered Inspectors Admitted under the Transitional Arrangement

A few number of Registered Inspectors were admitted by the Buildings Department via the special arrangement under s.3(7AA)(a) of the BO (Buildings Department, 2012c). Building professionals (RA, RPE and RPS of relevant disciplines/division of the ARB, ERB and SRB) of these kinds were "nominated" by their respective Registration Boards (the ARB, ERB and SRB) to become Registered Inspectors directly in the Inspectors' Registry of the Buildings Department. They did not necessarily hold any qualification(s) of AP and/or RSE.

Using the registration figures provided by Buildings Department (2012d), there was a total of 29 RIs were admitted by the Buildings Department under the Transitional Arrangement. Further studies revealed that out of these 29 RIs, 25 of them were in fact holding the qualifications of APs/RSEs; only 4 RIs (1 RI(A), 2 RI(E) and 1 RI(S)) did not hold the qualifications of AP and/or RSE.

7.3.2 Registered Inspector Admission that Does Not Require Professional Interviews

Under the qualifications prescribed in s.3(7AA)(b) of the BO, building professionals of RA, RPE and RPS holding the qualifications of AP or RSE may directly register themselves as RIs provided that they have gained experience in building repair and maintenance projects in Hong Kong within the seven years preceding the date of their applications (Buildings Department, 2011). Professionals from these groups (i.e., holders of AP or RSE in addition to RA, RPE or RPS) are not required to attend the professional interviews, nor vetted by the Inspectors Registration Committee (IRC).

Among the total of 308 RIs at 30 November 2012, RIs registered in the Inspectors' Registry of the Buildings Department were admitted as RI(A), RI(E) or RI(S) under this pathway.

7.3.3 Registered Inspector Admission that Requires Professional Interviews

Building professionals, such as RAs, RPEs and RPSs, who do not possess their additional qualifications as APs or RSEs, must fulfill the requirements set forth in the qualifications prescribed under s.3(7) of the BO (Buildings Department, 2011). In addition to having the required length of seven years of practical experience in building construction, repair and maintenance in Hong Kong, building professionals from these groups are required to attend professional interviews conducted by the IRC depending on the disciplines or divisions of the RA, RPE and RPS. In addition, their experiences are also required to be vetted by the IRC.

Out of the total 308 RIs registered in the Inspectors' Registry of the Buildings Department, only 49 RIs (16%) were admitted under this pathway. Such rate is unexpectedly low.

7.3.4 Summary on Registered Inspector Admission Rates

Summing up the three different pathways for RI registration in the Inspectors' Registry of the Buildings Department, the number of RIs registrations and their distributions are in respect of the three pathways, as shown in Table 7.1.

Out of the total 308 RIs registered in the Inspectors' Registry of the Buildings Department, 29 RIs (i.e., 8% of the total number) were admitted under the one-off Transitional Arrangement. The numbers of RIs admitted who were required or not required to attend professional interviews were 49 and 230 respectively. Therefore, the ratio of RIs admissions based on APs and RSEs and RAs, RPEs, and RPSs (i.e., required or not required to attend the professional interviews) is less than 1:4. This ratio signifies that four out of every five RIs admitted were not required to attend professional interviews because they were held APs and/or RSEs in addition to their professional qualifications of RA, RPE and/or RPS.

The above figures reveal that more than 80% of the admitted RIs were originated from the groups of AP or RSE. In other words, less than 20% of the RIs admitted were from building professionals of RA, RPE and/or RPS whom did not possess the qualifications of APs and/or RSEs.

Professional qualifications for all of the 308 RIs are detailed in Appendices 6.1 for RI(A), 6.2 for RI(E) and 6.3 for RI(S) respectively. However, their names are kept anonymous for privacy reason.

7.4 Evaluation of Registered Inspectors Application and Registration Rates

To evaluate the registration rate of RI, which is the percentage of the eligible building professionals that have been registered in the list of inspector registry holding by Buildings Department, the registration figures provided by Buildings Department (2012d) has been carefully reviewed in line with the membership databases collected by the local professional bodies including HKIA, HKIS and HKIE. The analysis can provide insight on the preferences of eligible professionals to rely themselves as RIs. As of 30 November 2012, a total number of around 8,455 building professionals (i.e., RA, RPE and RPS) were eligible to register themselves as RIs, subjected to prescribed qualifications detailed in s.3 of the BO (Buildings Department, 2011). After netting off the effect of multiple memberships in the relevant Registration Boards, the net number was reduced to 7,883. In addition, among the three RI-eligible building professionals, namely RA, RPE and RPS, there was a total number of 1,884 APs and RSEs whom were not required to attend the professional interviews conducted by the IRC, might register themselves as RIs. After netting off the effect of multiple memberships of AP or RSE, the number was reduced to 1,748.

The RI application and registration rates in comparison with the total populations of AP and RSE and RA, RPE and RPS are considered to be very low. Further analyses are made to evaluate the corresponding numbers or percentages in terms of the numbers of RI application and registration rates and the total populations of AP and RSE and RA, RPE and RPS.

Table 7.2
Evaluation of RI Application and Registration Rates under AP/RSE

Profession	AP/RSE	Authorised Persons / Registered Structural Engineers (Before netting Multiple Memberships) #5 Sub-Total (a)	Total (b)	(After netting Multiple Memberships) #5 Sub-Total (c)	Total (d)	RI(A)/RI(E)/ RI(S)	Registered Inspectors Applications and Registration Rates Application Rates (After Multiple Membership) #5 No. (e)	Percentage (f) = (e)/(d) x 100%	Registration Rates (After Multiple Membership) No. (g)	Percentage (h) = (g) / (d) x 100%
Architects	AP(A)	–	1,153	–	1152 #1	RI(A)	116	10%	94	8%
Engineers	AP(E)	140	551	6 #3	417 #4	RI(E)	101	24%	82	15%
	RSE	411		411						
Surveyors	AP(S)	–	180	–	179 #2	RI(S)	65	36%	54	30%
Total			1,884		1,748		282	16%	230	13%

Note #1: refers to 1 professional holding dual membership of AP(A) and RSE; and is included in thegroup of AP(E)/RSE (i.e. AP(A) = 1153 - 1 = 1152)

Note #2: refers to 1 professional holding triple membership of RSE, AP(E), and AP(S); and is included in the group of AP(E)/RSE (i.e. AP(S) = 180 - 1 = 179)

Note #3: refers to 133 professional holding dual membership of RSE and AP(E); and are included in the group of RE (i.e. RSE = 140 - 1 - 133 = 6)

Note #4: refers to the total AP(E) and RSE (i.e., AP(E)/ RSE = 140 + 411 - 133 - 1 = 417

Note #5: The number of Authorised Persons and registered structural engineers is captured at 30 November 2012 for calculating the registration rate using the figures provided by Buildings Department at 30 November 2012.

7.4.1 Registered Inspectors Application and Registration Rates under AP and RSE

Table 7.2 outlines the respective numbers of RI application and registration rates from the group of AP and RSE. According to the information provided by the Buildings Department (2012d), there were a total of 284 RI applications from the group of AP and RSE as of 30 November 2012. Taking out two withdrawn applications, the net number of RI applications from the group of AP and RSE was 282, which was equivalent to 16% of the net population of the group of AP and RSE. Among the 282 RI applications, 116 were AP(A), 101 were AP(E)/RSE and 65 were AP(S). These figures represent 10%, 24% and 36% of the net population of RI applications from the respective pool sizes of AP(A) (1,152), AP(E) and RSE (417) and AP(S) (179). Accordingly, although there were only 65 RI applications from the group of AP(S), AP(S) was ranked first among the three groups of AP and RSE. The group of AP(E) and RSE ranked second and AP(A) ranked third.

There were a total of 230 RIs whom registered themselves via the pathway of AP and RSE, representing 13% of the net population of the group AP and RSE. Whilst the largest number of RIs registration via the pathway of AP and RSE was AP(A) (i.e., 94), its percentage of registration (8%) was the lowest among the three groups of AP and RSE. Conversely, although there were only 54 AP(S) registered as RI(S), they were the most active group of RIs registrations and reached as high as 30% of the total number of AP(S). For AP(E) and RSE, there were a total of 82 successful RI registrations which was equivalent to 15% of the net population of AP(E) and RSE.

In Figure 7.1, it is worth noting that out of the 54 AP(S) whom registered themselves as RI(S), only three were originated from the Division of Quantity Surveying (i.e.,

Figure 7.1
RI(S) Eligible Professional Surveyors and Registration Rates via AP(S)

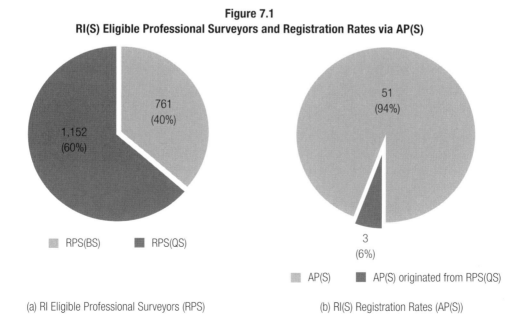

(a) RI Eligible Professional Surveyors (RPS)   (b) RI(S) Registration Rates (AP(S))

RPS(QS)), while the rest came from the Division of Building Surveying (i.e., RPS(BS)). The above figures signify that only less than 6% of the total number of RI(S) was originated from RPS(QS). The majority of RI(S) registrations were from RPS(BS), despite the fact that the number of RPS(QS) (1,152 members) was much higher than that of the RPS(BS) (761 members).

7.4.2 Registered Inspectors (RIs) Registration Rates under RA, RPE and RPS

Analyses reveal that RIs registration rates in the groups of RA, RPE and RPS (i.e., professional interviews conducted by the Inspectors Registration Committee are required) are far lower than that of AP and RSE for which professional interviews are not required. Based on the registration figures, only 49 out of the total 308 registered RIs were admitted under this pathway (i.e., after adjusting the 29 RIs admitted under the Transitional Arrangement and 230 RIs admitted via the pathway of AP and RSE). Among the 49 RIs admitted from the groups of RA, RPE and RPS, 29 were from RI(S), 18 were RI(E) and only two were RI(A). These figures do not correspond with the populations in the respective Registration Boards of ARB, ERB and SRB. Table 7.3 outlines the number of RIs registered via the pathway of RA, RPE and RPS and the corresponding populations in their Registration Boards.

Table 7.3 reveals the largest population of RI-eligible building professionals is ERB with a total member size of 3,660. Following are ARB and SRB with 2,882 and 1,913 registered members respectively. Mindful observation reveals a large number of multiple memberships existed in different disciplines and divisions of relevant Registration Boards. In fact, a large number of RAs, RPEs and RPSs were holders of APs and/or RSEs.

For the purpose of evaluating the registration rates under the groups of RA, RPE and RPS, it is again necessary to net off the number of multiple memberships from the Registration Boards and the number of AP/RSE from the total number of members in relevant Registration Boards.

Table 7.4 reveals that as of 30 November 2012, the net number of RI-eligible building professionals in relevant Registration Boards was reduced from 8,455 to 7,883 (column "d"). The Disciplines of RPE (Structural) and RPE (Civil) were found having the largest number of multiple memberships (424 numbers). Regarding AP and RSE, previous analysis found that 1,748 members were holding qualifications of APs and/or RSEs in addition to their professional qualifications of being RAs, RPEs and/or RPSs. The corresponding number of RAs, RPEs and RPSs after netting off the number of AP/RSE is then revised to 6,135 (column "h").

Based on the net number of RI-eligible building professionals (6,135) from respective Registration Boards (ARB, ERB and SRB) after adjusting the number of multiple

Table 7.3
RI-eligible Building Professionals and Numbers of RI Registrations
via RA, RPE and RPS Pathway

Registration Boards	Before netting off Multiple Memberships [5]		Registered Inspectors	RI Registration via RA/RPE/RPS	
ARB/ERB/SRB	Sub- Total	Total	RI(A)/RI(E)/RI(S)	Sub- Total	Total
ARB	–	2,882	RI(A)	–	2
ERB	RPE	3,660	RI(E)	–	18
Building	205		Building	7 [1]	
Structural	1,476		Structural	5 [2]	
Civil	1,871		Civil	4 [3]	
Materials (Building)	11		Materials (Building)	1	
Building Services (Building)	97		Building Services (Building)	1 [4]	
SRB		1,913	RI(S)		29
Building Surveying	761		Building Surveying	29	
Quantity Surveying	1,152		Quantity Surveying	0	
Total		8,455			49

Note [1]: refers to 1 RI holding dual RPE membership in Building and civil Disciplines. This member is included in PRE (Building).

Note [2]: refers to 1 RI holding dual RPE membership in Structural and Civil Disciplines. This member is included in RPE (Structural).

Note [3]: refers to 3 RI holding RPE (Civil) and dual RPE membership in Building, Structural and Building Services (Building) Discipline. 2 of the RIs included RPE (Building) and RPE (Structural Disciplines. The RI holding RPE (Civil) and Building Services (Building) is included in RPE (Civil)

Note [4]: refers to 1 RI holding dual RPE membership in Civil and Building Services (Building) Disciplines. This member is included in RPE (Civil) Discipline.

Note [5]: The number of professionals is captured at 30 November 2012.

memberships and the corresponding number of professionals holding additional professional qualifications of AP and/or RSE and the net number of RIs (49) registered in the Inspectors' Registry of the Buildings Department, the registration rates of individual Registration Boards as well as their disciplines and divisions can be evaluated. The results are tremendously different from that of the RI registration rates via the pathway of AP and RSE. The net RI application and registration rates via the pathway of RA, RPE and RPS, as of 30 November 2012 are detailed in Table 7.5.

Table 7.4
Net Number of RI-eligible Building Professionals RA, RPE and RPS

Registration Boards	Before netting off Multiple Memberships [#10]		After netting off Multiple Memberships [#10]		AP/RSE		Net RI Eligible Professionals (After Net of AP/RSE)	
	Sub-Total	Total	Sub-Total	Total	Sub-Total	Total	Sub-Total	Total
ARB/ERB/SRB	(a)	(b)	(c)	(d)	(e)	(f)	(g)	(h) = (d) - (f)
ARB	–	2,882	–	2,870 [#7]	–	1,152	–	1,718
ERB	RPE	3,660	RPE	3,178 [#8]	–	417	–	2,761
Building	205		165 [#1]		–		165	
Structural	1,476		1,264 [#2]		208 [#10]		1,056	
Civil	1,871		1,642 [#3]		209 [#10]		1,433	
Materials (Building)	11		10 [#4]		–		10	
Building Services (Building)	97		97		–			
SRB		1,913		1,835 [#9]	–	179	–	1,656
Building Surveying	761		730 [#5]		174		725	
Quantity Surveying	1,152		1,105 [#6]		5		931	
Total		8,455		7,883		1,748		6,135

Note [#1]: refers to 17 and 6 dual memberships with Civil and Structural Disciplines are included in Civil and Structural Disciplines respectively. Whereas 17 triple membership with Civil and Structural Disciplines are included in Structural Discipline (i.e., No. RPE (Building) = 205 - 17 - 6 - 17 = 165)

Note [#2]: refers to 6 dual memberships with Building Discipline and 17 triple membership with Building and Civil Disciplines are included in Structural Discipline. Whereas the 424 dual memberships with Civil Discipline are equally split between Structural and Civil Disciplines (i.e., NO. RPE (Structural) = 1,029 + 6 + 17 + 424/2 = 1,264)

Note [#3]: refers to 17 and 1 dual memberships with Building and Materials (Sublist) respectively are included in Civil Discipline. Whereas the 424 dual memberships with Structural Discipline are equally split (i.e., No. RPE (Civil) 1,412 + 1 + 17 + 424/2 = 1,642)

Note [#4]: refers to 1 dual membership with Civil Discipline are included in Civil Discipline (i.e., No. of RPE (Materials (Sublist) = 11 - 1 = 10)

Note [#5]: refers to out of 78 dual/triple memberships (1 + 12 + 65) between RPS and RPE/RA, there were a total of 31 members (0 + 6 + 25) from RPS (BS)

Note [#6]: refers to out of 78 dual/triple memberships (1 + 12 + 65) between RPS and RPE/RA, there were a total of 47 members (1 + 6 + 40) from RPS (QS)

Note [#7]: refers to 12 dual memberships with RPS are included in RA. 11 dual membership with RPE and 1 triple membership with RPS and RPE are included in RPE. (i.e., No. of RA = 2,882 - 1 - 11 = 2,870)

Note [#8]: refers to 65 and 11 dual memberships with RPS and RA respectively and 1 triple membership with RPS and RA. All are included in RPE. (i.e., No. of RPE = 3,101 + 65 + 1 + 11 = 3,178)

Note [#9]: refers to 65 dual memberships with RPS and 1 triple membership with RPE and RA are included in RPE. 12 Dual memberships with RA are included in RA (i.e., No. RPS = 1,913 - 65 - 1 - 12 = 1,835)

Note [#10]: refers to 417 members of AP/RSE from RPE (Civil) and RPE (Structural). Assume this figure is equally split between the Civil and Structural Disciplines

Note [#10]: The number of professional memberships are captured at 30 November 2012 (Buildings Department, 2012d).

Table 7.5
Net Registered Inspectors (RIs) Application and Registration Rate
under Pathways of RA/RPE/RPS

Registration Boards	Net RI Eligible Professionals (After Netting off of RA/RPE/RPS Multiple Membership and AP/RSE) [1]		RI Application Rates		RI Registrations via (RA/RPE/RPS)		RI Registration Rates (Percentage %)	
	Sub-Total	Total	No.	Percentage (%)	Sub-Total	Total	Sub-Total	Total
ARB/ERB/SRB	(a)	(b)	(c)	(d) = (c)/(b) x 100%	(d)	(e)	(f) = (d)/(a) x 100%	(g) = (e)/(b) x 100%
ARB	–	1,718	8	0.5%	–	2	–	0.1%
ERB	–	2,761	90	3.3%	–	18	–	0.7%
Building	165				7		4.2%	
Structural	1,056				5		0.5%	
Civil	1,433				4		0.3%	
Materials (Building)	10				1		10.0%	
Building Services (Building)	97				1		1.0%	
SRB	–	1,656	56	3.4%	–	29	–	1.8%
Building Surveying	725				29		4.0%	
Quantity Surveying	931				0		0%	
Total		6,135	154	2.5%		49		0.8%

Note [1]: The number of professional members is captured at 30 November 2012 (Buildings Department, 2012d).

The above results have revealed that the registration rates under the pathway of RA, RPE and RPS (i.e., professional interviews conducted by the IRC required) were surprisingly low. Evaluations of the results are as follows:

a) Registered Architects

There were only 2 RAs registered as RI(A) out of the total 1,718 eligible RAs and the registration rate was as low as 0.1%. According to Buildings Department (2012d), only 9 RAs have applied for RI(A). Taking out the only withdrawn application, the net number of RI applications was only 8, which was only equivalent to 0.5% of the total number of RA whom did not possess the additional qualification(s) of AP and/or RSE.

b) Registered Professional Engineers

The registration rate of RPE was slightly higher than that of RA's. Out of the total number of 2,761 RPEs, 18 were registered as RI(E). However, the registration rate of 0.7% is still considered to be very low. Likewise, there were a total of 90 RPEs (Buildings Department, 2012d), i.e., 3.3% of the total number of RPEs whom did not possess the additional qualifications of APs and/or RSEs. However, due to unavailable data in breaking down the applications of RPEs in different disciplines, the RI application rates in different RPE disciplines cannot be derived.

It is important to analyse the registration rates of different RPE disciplines. Table 7.6 indicates that Materials (Building) Discipline has the highest RI registration rate (10%) among all RPE disciplines with one registered member out of the total of 10. RPE Building Discipline ranked second with a registration rate of 4.2% (i.e., seven out of 165 members registered). Building Services (Building) Discipline ranked third with 1.0% registration rate, meaning only one out of the total of 97 members registered.

Surprisingly, the disciplines of Structural and Civil ranked penultimate and the last in RPE, with 0.5% and 0.3% registration rates respectively.

c) Registered Professional Surveyors (RPS)

According to the Buildings Department (2012d), 56 out of the total of 1,656 RPSs in Building Surveying Division (BSD) and Quantity Surveying Division (QSD) applied to be registered as RI(S). This figure represents 3.4% of the total population of corresponding RPS. Again, due to the unavailability of data in breaking down the applications of RPS in different BSD and QSD, the RI application rates in BSD and QSD cannot be derived.

Among the three RI-eligible building professionals which have to participate in professional interviews by the IRC, namely RA, RPE and RPS, RPS had the highest registration rate (1.8%). 29 out of the 1,656 RPSs were registered as RI(S). The registration rate was roughly 18 times and 2.5 times in comparison with that of the RA and RPE rates respectively.

It is of paramount importance to note that all the 29 RPSs were, in fact, coming from the RPS(BS) Division. The 29 RPS(BS) represent a total of 4% of the number of RPS(BS) whom do not possess the additional qualifications of APs and/or RSEs. More importantly, other than the three AP(S) originated from RPS(QS) who registered themselves as RI(S) as previously discussed, the records reveal that there was no RPS(QS) registered as RI(S) under the pathway of RPS at all, at least up to 30 November 2012.

Further studies reveal that the majority of professional building surveyors were in fact employed by the Government, tertiary institutions or utilities companies, etc. According to the Vocational Training Council (2012), there were as many as 76.6% (452 out of 590 respondents) of the professional building surveyors whom would not be available to serve as RIs for their employment reason. Based on this ratio, the total number of RPS(BS) employed by the Government, tertiary institutions or utilities companies etc. would be in the order of 593. The remaining number of RPS(BS) who might be available to serve as RI(S) would then be reduced significantly to only 168. Excluding the three RI(S) who came from RPS(QS) and AP(S), the net number of RI(S) originated from RPS(BS) was 90, out of the total of 168. Therefore, the registration rate of building surveyors, 53.6%, was astonishingly high when compared to the other 7 RI-eligible building professionals.

One of the major reasons of low RI registration rates in the seven RI-eligible building professionals other than building surveying should be resulted from the requirement of professional interviews set forth under PNAP APP-7. A study on the scope of the professional interviews conducted by the IRC will be discussed.

7.5 Scope of the Professional Interviews under PNAP APP-7

PNAP APP-7 (Buildings Department, 2011) stipulates that:

- s.2(b)—"a person may be included in the Inspectors' register without recommendation by the Inspectors Registration Committee (IRC) if the person is an AP or a RSE with experience as prescribed …";

- s.4(c)—"For RA, RPE in the building or structural engineering discipline or RPS in the building surveying division, he must, for a period or periods in aggregate of not less than one year within three years preceding the date of application, have practical experience in building construction, repair and maintenance gained in Hong Kong that the IRC considers appropriate;

- s.4(d)—"For RPE in the civil, building services (building) or materials (building) engineering discipline or RPS in the quantity surveying division, he must, for a period or periods in aggregate of not less than three years and of which at least one year falls within the three years preceding the date of application, have practical experience in building construction, repair and maintenance gained in Hong Kong that the IRC considers appropriate";

- s.6—"The function of the RC is to assist the BA in considering applications for inclusion in the relevant register. The RC will examine the qualifications and experience of the applicants and conduct professional interviews with the applicants"; and

- s.7—"The scope of professional interviews for AP, RSE, RGE and RI is detailed in Appendix A".

The above clauses clearly show that unless the applicant is an AP or a RSE, in addition to the prescribed qualifications of being an RA, RPE or RPS of relevant discipline and division, his or her experience will have to be vetted by the IRC and a professional interview will be required.

7.5.1 Appendix A of PNAP APP-7 for Professional Interviews for Registration as RI

Below are the relevant clauses extracted from Appendix A of PNAP APP-7 in relation to the professional interviews for registration as RIs (Buildings Department, 2011):

- Clause 5—"For the IRC, at least one member of the RC at a meeting hearing an application for inclusion should be of the same and at least another one should be of a different professional background as the applicant".

- Clause 6—"An applicant must satisfy the RC on his suitability for inclusion in the relevant register for which he applies. In this context, an applicant has to demonstrate that he has adequate practical experience and general knowledge in his profession to meet local requirements and to discharge his duties in Hong Kong. He will also be expected to have acquired a working knowledge of the BO and allied matters: the main criterion is a thorough understanding of general principles and fundamental requirements".

- Clause 7—"The principal subjects upon which the RC is likely to test an applicant's knowledge include:

  (a) The objectives of the BO and Regulations, and the mechanism of control; (Emphasis added)

  (b) The statutory role, functions and duties of an AP, RSE, RGE or RI as the case may be, and those of the BA in respect of private building development, building inspection and repair in Hong Kong;

  (c) Sufficient general awareness of local conditions to practice efficiently and effectively in Hong Kong without having to make frequent enquiries on matters of common local knowledge;

  (g) For registration as RI: A working knowledge of the BO and in particular those Regulations, relevant codes of practice, building safety and health standards pertinent to design and supervision of building works, building safety inspection, preparation of repairs or remedial works proposal and supervision of the necessary repairs or remedial works to remedy the defects or dilapidation of the building and/or to render the building safe; (Emphasis added)

  (h) The procedures for an AP, RSE, RGE or RI to follow in order to meet local statutory requirements; and

(i) Practice notes, circular letters and other advisory information published by government departments which are relevant to an AP, RSE, RGE or RI".

Obviously, the scope of the professional interviews for RIs not only focuses on the broad knowledge of the BO and Regulations and its mechanism of control, but also on relevant codes of practice, pertinent building safety and health standards. It includes design and supervision of building works, building safety inspection, preparation of repairs or remedial works, proposal and supervision of the necessary repairs or remedial works to remedy the defects or dilapidation of the building and/or to render the building safe. The professional interview for RIs demands candidates to have an extensive knowledge in (1) the BO and Regulations and relevant code of practice; (2) design, inspection, preparation of remedial proposal; and (3) supervision of building and remedial works in particular to building safety and health standards.

In light of the above, most RI-eligible building professionals, other than professional building surveyors, may not possess the breadth and depth of knowledge as required under Appendix A of the PNAP APP-7. This explains why application and registration rates across all seven RI-eligible building professionals are low except for the professional building surveyors.

7.6 Conclusions

Using the registration figures provided by local building authority at 30 November 2012, there were a total of 439 RI applications and 308 of them were admitted to be RIs. Out of the 308 admitted RIs, 29 were registered under the Transitional Arrangement; 230 via the pathway of APs and RSEs; and the rest via the pathways of RA, RPE and RSE. Whilst the overall RI admission rate was 78%, the admission rates under the pathways of RA, RPE and RPS were as low as 33% (2 out of 6), 26% (18 out of 69) and 59% (29 out of 49) respectively. Therefore, the overall admission rate under the pathways of RA, RPE and RPS (i.e., required to attend professional interviews) was only 40%.

Other than admission under AP(A), there were only two RAs registered as RI(A) out of the total number of 1,718 RI-eligible RAs and the registration rate was as low as 0.1%. Likewise, other than admission under AP(E) and RSE, there were only 18 RPEs from different RI-eligible ERB disciplines who registered as RI(E) out of the total 2,761 RPEs. Thus, the registration rate of 0.7% was also surprisingly low.

Among the 93 RI(S), other than three AP(S) and RPS(QS) who registered themselves as RI(S), there was no record showing any RPS(QS) had successfully registered themselves as RI(S) despite the fact that there were as many as 1,152 RPS(QS). Therefore, the net number of 90 RI(S) originated from RPS(BS) was astonishingly high when compared with seven other RI-eligible building professions.

The major reason of the low RI registration rates in the seven eligible building professions other than building surveying is because of the broad knowledge of building safety and health in relation to design, supervision and repairs required for the professional interviews set forth under PNAP APP-7.

References

Buildings Department (2011). *Registration of MBIS Registered Inspectors Begins*.Buildings Department of HKSAR. Available at http://www.bd.gov.hk/english/documents/news/20111229Bae.htm (Accessed on 21 March 2012)

Buildings Department (2012c). *Inspectors' Register*. Buildings Department of HKSAR. Available at http://www.bd.gov.hk/english/inform/index_ap.html (Accessed on 22 February 2015)

Buildings Department (2012d). *Planning and Lands Branch—Task Force on Building Safety and Preventive Maintenance as at 30 November 2012* (e-mail message to author, 6 December 2012). Hong Kong: Buildings Department of HKSAR.

Vocational Training Council (2012). *Building Inspection and Maintenance Training Centre*. VTC. Available at http://www.ive.edu.hk/ty/ivesite/html/en/campus/cn/ty_cn_ce.htm (Accessed on 28 November 2013)

CHAPTER 8

Projected Demand and Supply of Workforce on Registered Inspectors

8.1 Introduction

This chapter provides details on findings and analyses regarding the workforce condition of RIs, including:

(a) Key reason(s) causing RI-eligible respondents registering or not registering themselves as RIs;

(b) Under what scenario non-RI respondents might re-consider their registration as RIs;

(c) Key reason(s) causing the RI-eligible building professionals not planning to be RIs;

(d) The Key reason(s) for practising RIs not providing the building inspection and/ or repair services; and

(e) Workforce projection on the demand and supply of RIs under the implementation of the MBIS.

8.2 Data Collection

A two-stage web-based questionnaire survey was used to collect the perceptions of the building professional member and registered RIs and workforce data in relation to the inspection and rectification workers under the MBIS. Stage 1 was to (a) examine the number of RI-eligible building professionals whom had already registered as RIs and (b) to analyse the potential number of building professionals whom would register as RIs. Stage 2 was to assist the study on the number of RIs required for the implementation of the MBIS efficiently and provide forecast on RI workforce demand.

The questionnaire was administered to members by the professional institutions via their E-Broadcasts or E-Newsletters to their corporate members. Details of the questionnaire are shown in Appendix A.

Along with the databases of private buildings (in different building units group) analysed in Chapter 4, the estimated RI-time provided by respondents in Stage 2 would be used to estimate the RI-time that would be required for implementing the MBIS in the year of 2015. Projected RI-time (or the workforce demand) in the year 2025 can be developed through projected number of private buildings in different unit groups and RI-time required for inspection and rectification works under the MBIS (see Eq.[1]).

$$D_{RI} = f(B_i, D_{RI, inspection}, D_{RI, Rectification}) \dots\dots\dots\dots\dots\dots\dots\dots\dots\dots\dots\dots\dots Eq.[1]$$

where B_i = projected number of buildings in terms of number of units,

$D_{RI,\ inspection}$ = RI-time required for inspection works

$D_{RI,\ rectification}$ = RI-time required for rectification works

8.2.1   Analyses on Sources of Responses

a) Responses from Members of Professional Institutions

A total of 335 respondents completed the Stage 1 survey in mid-2012. 36 among all were not corporate members of the HKIA, HKIE or HKIS. They were not considered as "potential RI-eligible building professionals" under the MBIS and therefore discounted from further analysis. In this regard, only 299 valid responses were received.

Classifying by profession, engineers contributed the most respondents, with 151 valid responses (51%). Surveyors were the second, with a total of 129 valid responses (43%). Architects ranked the lowest, with only 19 valid responses (6%). Figure 8.1 illustrates the distribution of responses in different disciplines or divisions among the three professional institutions—the HKIA, HKIE and HKIS—and their weights in the total of 299 valid responses. Furthermore, categorising by relevant disciplines or divisions, the Building Surveying division comprised the largest proportion of respondents, with 66 valid responses (22%); the second was the Quantity Surveying division, with 63 (21%); the third was the Civil discipline, with 50 (17%), and the last was the Materials discipline, with only 6 (2%).

Figure 8.1
Response Rates of Potential RI-eligible Building Professionals
(Individual Disciplines or Divisions of the HKIA, HKIE and HKIS)

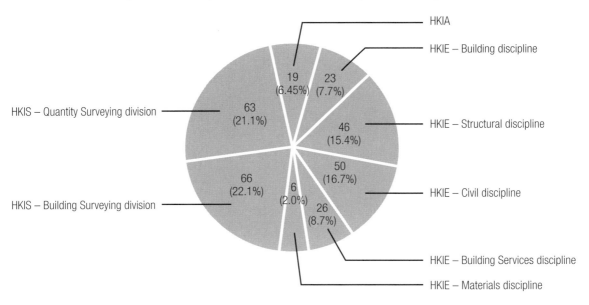

Figure 8.2
Response Rates of RI-eligible Building Professionals

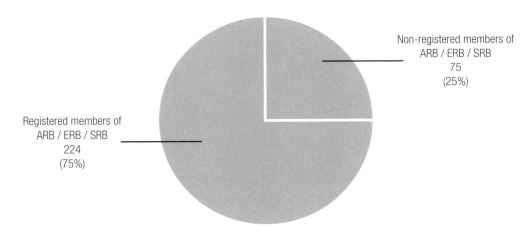

Figure 8.3
Response Rates of RI-eligible Building Professionals
(Individual Disciplines or Divisions of the ARB, ERB and SRB)

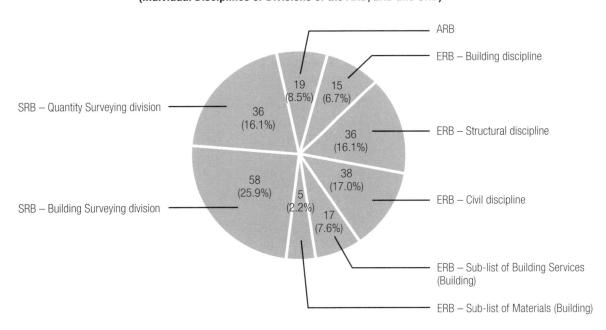

b) Responses from Registered Members of Registration Boards

Among the 299 valid responses, 224 (75%) were registered members in relevant disciplines or divisions of the ARB, ERB and SRB and 75 (25%) were not (Figure 8.2). The 224 registered members were regarded as RI-eligible building professionals under the MBIS.

Figure 8.3 shows the total of 224 responses from registered members of the ARB, ERB and SRB. Ranking in terms of number of responses are very similar to that of the HKIA, HKIE and HKIS. By profession, engineers comprised the largest number of respondents, with 111 valid responses (50%). Surveyors were the second with 94 valid responses (42%). The architects ranked the lowest, with only 19 valid responses (8%).

However, by relevant disciplines and divisions, the ranking was to an extent different from that of the HKIA, HKIE and HKIS. While the largest number of respondents was still from the Building Surveying division, with 58 (26%) valid responses, the second was from the Civil discipline, with 38 (17%) valid responses, instead of the Quantity Surveying division. The Quantity Surveying and Structural disciplines both ranked the third with 36 valid responses (16%). The Materials discipline ranked the lowest, with 5 valid responses (2%).

c) Response Rate from Registered Inspectors (RIs)

Among the 299 valid responses, 52 RIs responded to the Stage 1 survey. This figure, which is an equivalence of 16.9% of the total number of responses, shows highly encouraging responses from the total pool of 308 RIs (as of 30 November 2012) listed in the Inspectors' Registry of the Buildings Department.

Table 8.1 tabulates the numbers and percentages of responses in different RI disciplines. It is reckoned that questionnaire from 21 RI(S) out of the total 93 RI(S) (22.6%); 20 out of the total 109 RI(E) (18.3%); 11 out of the total pool of 106 RI(A) (10.4%) have been received.

Table 8.1
Statistics of Responses from RIs

	Inspectors' Registry	No. of RI	Responses in RI-Disciplines	
			No.	%
1	Registered Inspectors (Architects)(RI(A))	106	11	10.4%
2	Registered Inspectors (Engineers)(RI(E))	109	20	18.3%
3	Registered Inspectors (Surveyors)(RI(S))	93	21	22.6%
	Total :	308	52	16.9%

8.2.2 Backgrounds of Respondents

a) Nature of Organisations

The respondents of the Stage 1 survey were come from various sectors/disciplines of the construction industry. 61 (20.4%) were from consulting firms (new works); 53 (17.7%) from contracting firms (new works); and 9 (3%) from tertiary institutions or utility companies. Figure 8.4 shows the distribution of employing organisations in the construction industry of the respondents.

There were 25 (8.4%) and 16 (5.4%) respondents coming from consulting firms (decoration or maintenance) and contracting firms (decoration or maintenance) respectively, making up 13.8% of the total responses.

Figure 8.4
Analysis of Respondents (Nature of Organisation/discipline)

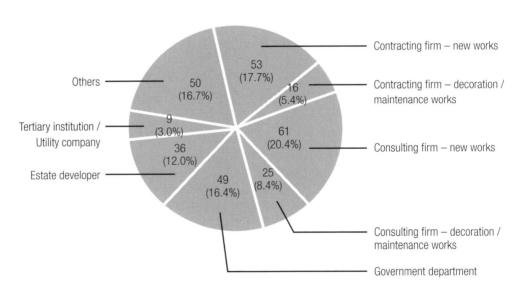

b) Positions Held by the Respondents

Professional architects, engineers or surveyors comprised the largest sector of respondents (33.4%). A significant proportion of respondents came from senior level position in their organisations, with 64 respondents (21.4%) holding positions as Director or Chief Executive Officer. Figure 8.5 shows the distribution of respondents' positions.

Figure 8.5
Analysis of Respondents (Positions in Construction Industry)

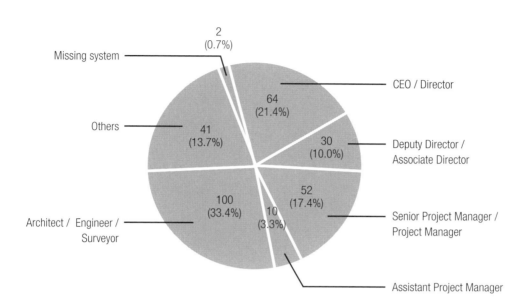

c) *Construction-related Work Experiences*

The majority of respondents had extensive experience in the construction industry (Figure 8.6). Among the 299 valid responses, 156 respondents (52.2%) had 21 years and above construction-related experience. The other two sectors of respondents (11 to 15 years and 16 to 20 years) were also considered to be very experienced in the field. Coincidently, there were 55 respondents (18.4%) in each of these two sectors. Therefore, there were 266 respondents (89.0%) out of the total 299 valid responses holding 11 years or more construction-related work experiences.

d) *Maintenance-related Work Experiences*

Unlike for construction-related work experience, the respondents had varying levels of experience in maintenance-related experience (Figure 8.7). The largest sector of respondents (28.8%) held 1 to 5 years maintenance-related work experience. 32 respondents (10.7%) held 21 years and above experience. 42 respondents (14.0%) only had less than a year of maintenance-related work experience.

Figure 8.6
Construction-related Work Experience of the Respondents

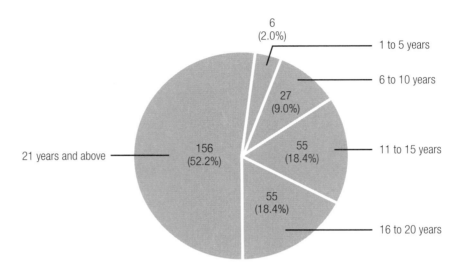

Figure 8.7
Maintenance-related Work Experiences of the Respondents

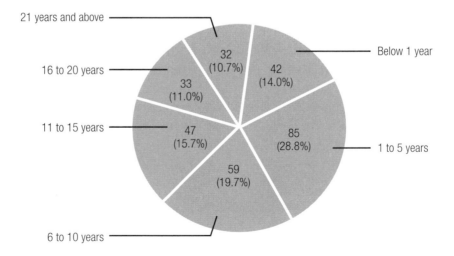

8.3 Evaluation of the Numbers of Aged Buildings under MBIS in 2015 and 2025

Under the MBIS, 2,000 target buildings are selected each year by the Buildings Department and the buildings selected would represent a mix in different conditions and age profiles in different districts (Buildings Department, 2013c). Using the database of private buildings as analysed in Chapters 4 and 5, the 2,000 private buildings in different unit groups can be estimated.

In addition, with the data collected from Stage 2 respondents in which the estimated RI-time for inspection and/or repair works of different unit groups of private buildings are provided, the estimated total number of RIs required for the implementation of the MBIS can be predicted.

Given the increase in building unit size since the 1970s due to the advancement of construction technology and high land cost, the size of aged buildings (in terms of number of building units) increased as a result. This means that more RI-time would be needed to carry out the necessary inspection and/or repair works due to a greater share in larger-sized private buildings (in terms of number of building units) under the MBIS.

The estimation process of RI-time required by the MBIS (or the workforce demand of registered inspector) in 2015 and 2025 is shown in Figure 8.8. Accordingly, the numbers of private buildings per unit group falling under the MBIS in 2015 and 2025 should be based on the study of age distributions of buildings at the end of 2014 and 2024 which are shown in Table 8.2.

Under the MBIS, the number of different building unit groups will change with time, even though the total number of private buildings being inspected would be the same, i.e., 2,000 buildings per year. As shown in Table 8.2, for the years 2014 between 2024, the number of private buildings aged 30 years or above will increase by 4,713. This is approximately an increase of 471 per year. As shown in Table 8.2, the medium-sized and large-sized buildings which will be fallen under the MBIS has been increased by 44.8% and 155.1% which compared with the figures captured at 31 December 2014.

Further analyses would be conducted as follows:

(a) The estimated number of RIs required based on the data collected from all respondents, i.e., RI respondents and Non-RI respondents;

(b) The estimated number of RIs required based on the data collected from different professional disciplines, i.e., RI(A), RI(E) and RI(S); and

(c) The estimated number of RIs required based on the data collected from practising RI respondents and non-practising RI respondents.

Figure 8.8
Flow Chart of the RI-time Estimation Process under MBIS

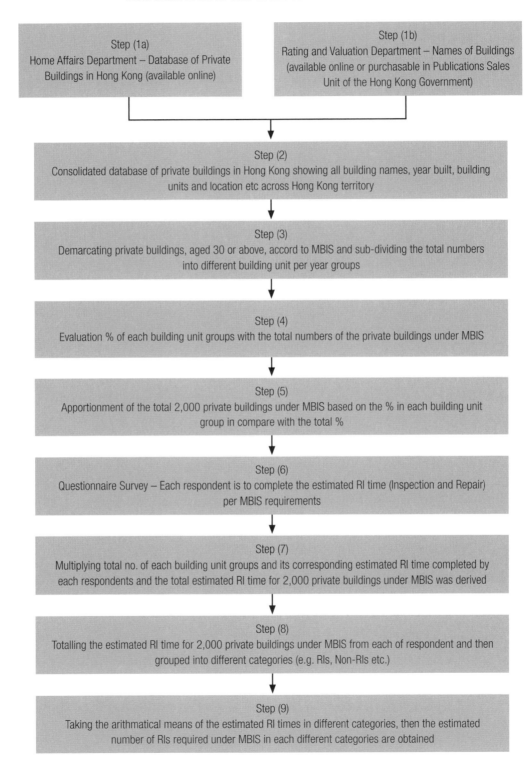

Step (1a)
Home Affairs Department – Database of Private Buildings in Hong Kong (available online)

Step (1b)
Rating and Valuation Department – Names of Buildings (available online or purchasable in Publications Sales Unit of the Hong Kong Government)

Step (2)
Consolidated database of private buildings in Hong Kong showing all building names, year built, building units and location etc across Hong Kong territory

Step (3)
Demarcating private buildings, aged 30 or above, accord to MBIS and sub-dividing the total numbers into different building unit per year groups

Step (4)
Evaluation % of each building unit groups with the total numbers of the private buildings under MBIS

Step (5)
Apportionment of the total 2,000 private buildings under MBIS based on the % in each building unit group in compare with the total %

Step (6)
Questionnaire Survey – Each respondent is to complete the estimated RI time (Inspection and Repair) per MBIS requirements

Step (7)
Multiplying total no. of each building unit groups and its corresponding estimated RI time completed by each respondents and the total estimated RI time for 2,000 private buildings under MBIS was derived

Step (8)
Totalling the estimated RI time for 2,000 private buildings under MBIS from each of respondent and then grouped into different categories (e.g. RIs, Non-RIs etc.)

Step (9)
Taking the arithmatical means of the estimated RI times in different categories, then the estimated number of RIs required under MBIS in each different categories are obtained

Table 8.2
Projected Numbers of Private Buildings (Per Building Unit Groups)
to be Selected by the Buildings Department

Year	Type	No. of Private Buildings (Per Unit Groups)					Remarks
		50 or below	51 to 200	201 to 400	401 and above		
		Small	Medium	Large	Extra Large	Total	
Dec - 14	Total No. of Buildings (Per Unit Group)	15,350	5,626	2,424	397	23,797	
	Overall % in HK (Per Unit Groups)	64.5%	23.6%	10.2%	1.7%	100.0%	
Dec - 14	Total No. of Private Buildings under MBIS Year 2014	11,831	3,098	563	89	15,581	Private Buildings completed on or before 1984
		75.9%	19.9%	3.6%	0.6%		
	Estimated Apportioned No. of Buildings being selected by BD (2014)	1,518	398	72	12	2,000	Based on apportioned buildings completed or before 1984
Dec - 24	Projected Total No. of Private Buildings under MBIS by Year 2024	14,094	4,487	1,436	277	20,294	Private Buildings completed on or before 1994
		69.4%	22.1%	7.1%	1.4%		
	Estimated Apportioned No. of Buildings being selected by BD (2024)	1,388	442	142	28	2,000	Based on apportioned buildings completed on or before 1994
Estimated Movements	No. of Buildings	2,263	1,389	873	188	4,808	
	Grown rate	19.1%	44.8%	155.1%	211.2%	30.2%	

Note: For Private Buildings Non-Domestic and Domestic Buildings over 3 storeys.

8.4 Workforce Demand and Supply under MBIS

In Stage 2 of the survey, a total of 35 registered RIs responded to the questionnaire (Table 8.3). This represents 11.4% of the total 308 RIs as of 30 November 2012. In addition, 22 of the 35 RIs (63%) were practising in the field of building inspection and/or repair works. Table 8.3 shows the distribution of the 35 RI respondents of Stage 2 of the survey, in terms of building professions and practising in the field of building inspection and/or repair works.

Table 8.3
Summary of the Respondents at Stage 2 Questionnaire (Practising/Non-Practising RIs)

Registered Inspectors (RI)	Practising Professionals	Non-Practising Professionals	Total
RI(A)	6	3	9
RI (E)	5	7	12
RI (S)	11	3	14
Total :	22	13	35

Notes
1. "Practising Professionals" donates the professionals who are being involved in the field of building inspection and/or repair works
2. "Non-Practising Professionals" donates the professionals who are not being involved in the field of building inspection and /or repair works

8.4.1 Estimated Workforce Demand on Registered Inspectors (RIs)

Stage 2 of the survey was set with the objective of estimating the number of RIs (or workforce demand) that would be needed under the MBIS. The respondents were asked to provide their estimated RI-time that would be required for implementing the inspection and/or repair works for private buildings of different sizes, according to the Code of Practice (CoP) for the MBIS and MWIS published by the Buildings Department (Buildings Department, 2012a, 2012b).

To evaluate the workforce demand of RIs, the labour multiplier approach is adopted. It makes use of linear correlation between the workload (e.g., number of buildings to be inspected) and workforce demand per building (Sing et al., 2012, 2014). Determining the number of working days per year is an issue that needs to be addressed. In the case of Hong Kong, it is assumed that a building professional works 22 days per month. With the adjustment on statutory and public holiday, it is estimated that s/he will work 264 days per year. This assumption forms the basis of evaluation for the demand for RI per year for the implementation of the MBIS. In the survey, RI(A), RI(E), or RI(S) were asked to provide: (a) the number of hours required for inspection ($H_{inspection, p, i}$) and the number of hours required for rectification works ($H_{rectification, p, i}$), of private buildings under the MBIS. Their relationship is expressed as:

(a) Average workforce demand for Registered Inspectors required for the inspection of buildings under the MBIS:

$$D_{RI, inspection} = \frac{\sum_{i=1}^{4} \sum_{p=1}^{3} B_i \cdot H_{inspection, p, i}}{3n} \dots\dots\dots\dots\dots\dots\dots\dots\dots\dots\dots\dots\dots\dots Eq.[2]$$

where

B_i = projected number of buildings in terms of number of units

i = group of buildings with different number of units, e.g., $i = 1$ for small-sized buildings

$H_{inspection, p, i}$ = numbers of hours required for inspection based on the professional discipline (p) and group of the buildings (i)

p = professional from different disciplines, e.g., 1 = Registered Inspector (Architect), 2 = Registered Inspector (Engineer) and 3 = Registered Inspector (Surveyor)

n = number of working days per year

(b) Workforce demand of Registered Inspectors required for rectification of defects in the building under the MBIS

$$D_{RI, rectification} = \frac{\sum_{i=1}^{4} \sum_{p=1}^{3} B_i \cdot H_{rectification, p, i}}{3n} \dots\dots\dots\dots\dots\dots\dots\dots\dots\dots\dots\dots\dots\dots Eq.[3]$$

where

B_i = projected number of buildings in terms of number of units

i = group of buildings with different number of units, e.g., i = 1 for small-sized buildings

$H_{rectification, p, i}$ = numbers of hours required for rectification works based on the professional discipline (p), and group of the buildings (i)

p = professional from different disciplines/divisions, e.g., 1 = Registered Inspector (Architect), 2 = Registered Inspector (Engineer) and 3 = Registered Inspector (Surveyor)

n = number of working days per year

The workforce demand of RI (i.e., number of RI per year) can be estimated with Eqs. [2] and [3]. Table 8.4 provides the average RI workforce demand based on the RI-time as reported by RIs from different professional disciplines.

(a) From the point of view of RI(A)

- Inspection — 95.04 RIs (2015), (106.39 RIs, 2025)
- Rectification and Repair — 361.24 RIs (2015), (381.86 RIs, 2025)
- Total — 456.28 RIs (2015), (488.25 RIs, 2025)

Table 8.4
Estimated Workforce Demand of RIs under MBIS (Year 2015 and 2025)

Respondent		RI(A)	RI(E)	RI(S)	RI Total	RI Practising	RI Non Practising
No.		9	12	14	35	22	13
2015 Inspection	RI No. Required	95.04	94.79	80.98	89.33	78.93	106.93
	Variance (%)	6.4%	6.1%	-9.3%	0.0%	-11.6%	19.7%
Rectification	RI No. Required	361.24	352.26	344.04	351.28	316.83	409.57
	Variance (%)	2.8%	0.3%	-2.1%	0.0%	-9.8%	16.6%
Total	RI No. Required	456.28	447.05	425.02	440.61	395.76	516.51
	Variance (%)	3.6%	1.5%	-3.5%	0.0%	-10.2%	17.2%
2025 Inspection	RI No. Required	106.39	105.05	89.45	99.15	87.31	119.19
	Variance (%)	7.3%	5.9%	-9.8%	0.0%	-11.9%	20.2%
Rectification	RI No. Required	381.86	373.45	361.11	370.68	335.20	430.71
	Variance (%)	3.0%	0.7%	-2.6%	0.0%	-9.6%	16.2%
Total	RI No. Required	488.25	478.51	450.55	469.83	422.52	549.90
	Variance (%)	3.9%	1.8%	-4.1%	0.0%	-10.1%	17.0%

(b) From the point of view of RI(E)

- Inspection — 94.79 RIs (2015), (105.05 RIs, 2025)

- Rectification and Repair — 352.26 RIs (2015), (373.45 RIs, 2025)

- Total — 447.05 RIs (2015), (478.51 RIs, 2025)

(c) From the point of view of RI(S)

- Inspection — 80.98 RIs (2015), (89.45 RIs, 2025)

- Rectification and Repair — 344.04 RIs (2015), (361.11 RIs, 2025)

- Total — 425.02 RIs (2015), (450.55 RIs, 2025)

(d) From the point of view of practising RIs

- Inspection — 78.93 RIs (2015), (87.31 RIs, 2025)

- Rectification and Repair — 316.83 RIs (2015), (335.20 RIs, 2025)

- Total — 395.76 RIs (2015), (422.52 RIs, 2025)

(e) From the point of view of non-practising RIs

- Inspection — 106.93 RIs (2015), (119.19 RIs, 2025)

- Rectification and Repair — 409.57 RIs (2015), (430.71 RIs, 2025)

- Total — 516.51 RIs (2015), (549.9 RIs, 2025)

Summing up the above, the average workforce demand for full-time RIs in 2015, as estimated by RIs, would be as follows:

- Inspection — 89.33 RIs

- Rectification and Repair — 351.28 RIs

- Total — 440.61 RIs (≈ 441 RIs at year 2015)

Likewise, the average workforce demand for full time RIs in the next 10 years (i.e., till 2025), as estimated by RIs, would be as follows:

- Inspection — 99.15 RIs

- Rectification and Repair — 370.68 RIs

- Total — 469.83 RIs (≈ 470 RIs at year 2025)

Stage 2 survey reveals that around 89.33 full time RIs would be required for inspection of target buildings in 2015. However, if inspection and repair are to be implemented at the same time, another 351.28 RIs would be required. Therefore, the total number of RIs required would be around 440.61 when the MBIS is implemented at full speed and in all aspects in 2015.

Comparing the workforce demand for RIs in 2015 and 2025, it is estimated that the total number of full time RIs that required to inspect and repair a total of 2,000 buildings under the MBIS would increase from 440.61 to 469.83. This movement represents an increase of 6.6% of RI time due to the increase in size (in terms of building units) of private buildings from 2015 to 2025.

Among the three groups of RIs, it is noted that RI(S) had the lowest estimated RI-time for both inspection and/or rectification stages. Their total estimated RI-time was approximately 3.5% (-15.59 number of RIs) below the average estimated RI-time. The estimated RI-times provided by RI(A) and RI(E) were 3.6% (+15.67 number of RIs) and 1.5% (+6.44 number of RIs) above that of the average estimated RI-time.

It is of paramount importance to note that the estimated RI-time as considered by practising professionals were of 10.2% (-44.85 number of RIs) below the average estimated RI-time. For non-practising professionals, they considered that the estimated RI-time was around 17.2% (+75.9 number of RIs) higher than that of the average estimated RI-time.

Comparing the estimated RI-times in 2015 and 2025, with the assumption that the same number, i.e., 2,000 buildings are to be inspected and rectified or repaired according to the MBIS, it is estimated that the overall RI-time will increase by 29.22 RIs (i.e., from 440.61 to 469.83 RIs) in the next 10 years.

8.4.2 Estimated Supply of RIs for Year 2015

To ensure fair competition, the Government considered that the market should have a supply of at least 300 RIs when the first prescribed inspection under the MBIS commenced in early 2013 (Development Bureau, 2011c; Legislative Council, 2011b, 2011d, 2011e, 2012a). However, Hon Kam Nai-wai, a member of the Legislative Council, opined that the supply of 300 RIs for the first prescribed inspection was inadequate and may have a negative effect on the inspection cost (Legislative Council, 2012b).

As of 22 February 2015, a total of 420 RIs were listed in the Inspectors' Registry of the Buildings Department, closely matching the initial expectation of the Legislative Council. However, the survey results reveal that quite a number of the RIs were, in fact, not practising. Based on the results from Stage 1 of the survey, only 57.7% were practising RI.

Based on this ratio, the estimated number of practising RIs out of the total 420 RIs listed in the Inspectors' Registry of the Buildings Department as of 22 February 2015, is calculated as follows:

No. or RIs as at 22 February 2015

By the same token and based on the same logic, the estimated number of non-practising RIs out of the 308 RIs is:

No. or RIs as at 22 February 2015

To sum up, it is safe to assume that the number of practising RIs, as of 22 February 2015, would be 242 (57.7%) and non-practising RIs would be 178 (42.3%).

8.4.3 Evaluation of the Professional Workforce

The analyses reveal that, depending on the groups of building professionals, the number of RIs required for the MBIS to be implemented at full speed and in all aspects in 2015 is 441. On the other hand, although there were a total of 420 RIs listed in the Inspectors' Registry of the Buildings Department, the estimated number of practising RIs was 242 only.

Taking the 441 RIs required under the MBIS and 242 listed practising RIs in the field of building inspection and/or repair works into consideration, the shortage of RIs to smoothly and efficiently implement the MBIS in 2015 would be 199, which was equivalent to 82% (i.e., 199/242 x 100%) of the total number of RIs listed in the Inspectors' Registry of the Buildings Department.

In connection with the supply of RIs, the suggestion made by Dr. Hon Raymond Ho Chung-tai (Legislative Council, 2011c) to expand the pool of RIs with other experienced technical personnel like associate members of the HKIE and affiliate members of the HKIA was not promoted. Some members of the Legislative Council also suggested creating training programmes on the essential skill-set and knowledge for RI registration for personnel undertaking works required under the MBIS. None of the three professional institutions (the HKIA, HKIE and HKIS) had plans to organise top-up courses to train their associate or affiliate members to become RIs. The HKIS stressed that the duties of RIs should be performed by "professionally-qualified" personnel and considered that top-up courses would not be adequate to equip a technical practitioner with the knowledge to conduct prescribed inspections or supervise prescribed repairs under the MBIS. In this regard, further research was made to examine the pools and size of associate or affiliate members, i.e., technical members, of the three professional institutions.

As shown in Table 8.5, there were a total of 647 technical members in the relevant disciplines or divisions of the HKIA, HKIE and HKIS. This figure represents only a very small portion of the total corporate members of the institutions. Balancing the number of technical members, the training resources that needed to top-up technical members to professional RIs, and the professional skills and knowledge in a number of areas as required by the MBIS, the advice from the HKIA, HKIE and HKIS for not providing the top-up training course is shared.

Despite the shortage of RIs as analysed above, the views from the HKIS (HKIS, 2010, p. 5) stated that "the government should stress on the expertise of the RIs, instead of their numbers .... If such qualification threshold is relaxed merely for the sake of expanding the pool of RIs, it will allow incompetent persons to carry building inspections, and handling the maintenance works at the expense of the quality of inspections and repairs, i.e., building safety, hence defeating the purpose of the MBIS".

Table 8.5
Statistics of Technical Members of the HKIA, HKIE and HKIS

	Professional Institution	Affiliate Member	Associate Member	Total
1	Hong Kong Institute of Architects (HKIA) [#1]	15	33	48
2	Hong Kong Institute of Engineers (HKIE) [#2]			588
	a. Building discipline	N/A	100	
	b. Structural discipline	N/A	34	
	c. Civil discipline	N/A	204	
	d. Materials discipline	N/A	6	
	e. Building Services discipline	N/A	244	
3	Kong Institute of Surveyors (HKIS) [#3]			11
	a. Building Surveying division	N/A	0	
	b. Quantity Surveying division	N/A	11	
	Total :	15	632	647

Note [#1]: Numbers of members obtained from HKIA website "Looking for architects" available at gttp://www.hkia.net/en/LookingForArchitects/LookingForArchitects_02.htm(Accessed on 1 March 2013) (HKIA, 2013)

Note [#2]: Numbers of members as advised by email from the Qualification and Membership Board of the Hong Kong Institution of Engineers on 12 March 2013 (HKIE, 2013)

Note [#3]: Numbers of members obtained from HKIS website "Membership Statistics as at 1 January 2013" available at http://www.hkis.org.hk/ufiles/memstat-20130101.pdf (Accessed on 1 March 2013) (HKIS, 2013)

8.5 Perceptions of Eligible RIs towards MBIS

Under the MBIS, the following building professionals are considered to have obtained prescribed qualifications for the purpose of registering as RIs (s.4(a) of PNAP APP-7) (Buildings Department, 2011):

(a) Authorised Persons (AP(A), AP(E) and AP(S));

(b) Registered Structural Engineers (RSE);

(c) Registered Architects (RA);

(d) Registered Professional Engineers (RPE) in building, structural, civil, building services (building) and materials (buildings);

(e) Registered Professional Surveyors (RPS) in building surveying or quantity surveying divisions.

The previous analyses reveal that, as of 22 February 2015. There were a total of 17,600 building professionals registered in the 8 relevant disciplines or divisions of the HKIA, HKIE and HKIS, with a total of 9,385 professional members registering themselves in the relevant registration boards (the ARB, ERB and SRB) (Note: The net number of RI-eligible building professionals was 8,650 after netting off the effect of multiple registrations in different registration boards). These professionals may be eligible to register themselves as RIs, subject to the prescribed requirements set forth under PNAP APP-7.

Stage 1 of the survey finds out that a large number of potential RI-eligible building professionals were in fact interested and planned to register as RIs. 124 out of the 299 valid responses of Stage 1 of the survey expressed interests in RI registration (See Table 8.6).

Table 8.6
Projected No. of Building Professionals
Who Expressed Interests in RI Registration

Item	Respondents' Planning to Register as Register Inspectors	Professional Institutions/ Registration Boards	No. of Respondents (Note [1])	Percentage (%)	Projected No. of Professionals (Note [2])
1	Plan to register in the List of Inspectors' Registry with one (1) year	ARB/ERB/SRB [3]	33	11.0%	952
		HKIA/HKIE/HKIS [4]	40	13.3%	2,340
2	Plan to register in the List of Inspectors' Registry with three (3) years	ARB/ERB/SRB [3]	7	2.3%	199
		HKIA/HKIE/HKIS [4]	13	4.3%	757
3	Plan to register in the List of Inspectors' Registry but timing of registration is not yet fixed	ARB/ERB/SRB [3]	55	18.4%	1,592
		HKIA/HKIE/HKIS [4]	71	23.7%	4,172
	Total number of building professionals who plan to register in the List of Inspector Registry:	ARB/ERB/SRB [3]	95	31.8%	2,751
		HKIA/HKIE/HKIS [4]	124	41.3%	7,269

Note [1]: donates the number of respondents from the Stage 2 questionnaire survey respondents

Note [2]: donates the "projected number" of the potential RI eligible professionals based on the ratio (percentage) of Stage 2 respondents and total number of corporate/ registered members in respective registration boards and professional institutions

Note [3]: total number of registered members in ARB/ERB/SRB = 8,650 (after netting the effect of multi-professional registration and as of 22 February 2015)

Note [4]: total number of corporate members in HKIA/HKIE/HKIS = 17,600 (as of 22 February 2015)

Based on the total number of 8,650 RI-eligible building professionals registered in the ARB, ERB and SRB (after netting the effects of multiple professional registration and as of 22 February 2015), the "projected number" of RI-eligible building professionals whom were interested or planning to be listed in the Inspectors' Registry of the Buildings Department was 2,751. If the projection is based on the total number of potential RI-eligible corporate members in the professional institutions of the HKIA, HKIE and HKIS, this figure could reach as many as 7269. Regardless, these two figures fully support the advice given by the HKIA, HKIE and HKIS that there should be adequate professionals planning to register as RIs as supported by the enthusiastic participation and responses of their members during the discussion on the MBIS (Development Bureau, 2011d; Legislative Council, 2011a, 2011e).

Putting aside the chosen methods first, the projected number of professionals interested or planning to be listed in the Inspectors' Registry was 2,751 or 7269; these figures are far higher than the actual number of 420 registered RIs, as of 22 February 2015.

Based on the results from Stage 1 of the survey, only 41 building professionals out of the total of 299 valid responses, regardless of whether or not they were RIs, responded that they were participating in the field of building inspection and/or repair works (see Figure 8.5). This figure more or less signifies 14% of the total population of potential RI-eligible building professionals who are serving the field of building inspection and/or repair works.

In the past, there were a very limited number of professional practitioners "dominating" the field of building inspection and/or repair works. Given that the nature of the practice and/or repair works of building inspection is quite different from new building works, most of the professional practitioners are AP(A), AP(S), registered or professional architects and building surveyors. They are largely involved in the field of building inspection and/or repair works as they are more familiar with the requirements and procedures set forth. There are also a limited number of RSEs or registered or professional structural engineers who assist with the functions of structural design or works in the course of building inspection and/or repair works. They seldom act in the roles of leading consultants but as team players in building inspection and/or repair works. Moreover, the professional practitioners of building inspection and/or repair works and new building works seldom shift their jobs from one to another.

However, the demand of this "unique market" of building inspection and/or repair works suddenly increased tremendously following the newly legislated MBIS. Such increase would inevitably attract building professional practitioners of new building works to consider shifting their roles or to extend their horizons to building inspection and/or repair works. It is strongly believed that this is one of the key, if not the sole reason, that a great number of building professionals have expressed their interest to participate in building inspection and/or repair works.

8.6 Reasons that RI-eligible Building Professional Do Not Plan to Be Listed as RIs

From Stage 1 of the survey, there was a number of RI-eligible building professionals that have not registered as RIs nor did they have any plans registering. To explore the reasons behind, questions were set out to respondents. By so doing, it is considered that the Government, the policy makers and the relevant authorities or professional institutions may take appropriate action(s) to promote RI registration.

124 potential RI-eligible building professionals advised that they were planning to register as RIs (Table 8.6). On the other hand, another 119 provided their reasons for not registering. Taking out 8 respondents (Group 1) who did not answer this question, there were a total of 111 respondents who provided their reasons. The results are summarised in Table 8.7.

Based on Table 8.7, 36 respondents (Group 2) advised that they have no interest in registering as RIs. This amounts to around one third (32.4%) of the total respondents. Another 16 respondents (Group 3) advised that they prefer working in other construction-related professional services as opposed to RI. This amounts to 14.4% of the total respondents.

It is important to note that there were another 16 respondents (14.4%) (Group 4) who considered the responsibilities and liabilities of being an RI under the MBIS unclearly set. There were also eight respondents (7.2%) (Group 5) who considered the current remuneration of being an RI was not corresponding to the responsibilities and liabilities set forth in the MBIS. Mindful observation further reveals that another eight respondents (7.2%) (Group 7) considered that the word "inspector" misconceived as "para-professional", leading to low social recognition. Four respondents (3.6%) (Group 6) considered the opportunity cost of providing RI services was higher than other construction-related professional services.

From the above, it was considered that 52 (46.8%) respondents from Groups 2 and 3 can be regarded as having no interest in serving the field of RI unless there is a drastic change or reform in the construction industry of Hong Kong. There were also two other "useful hints" indicated by the respondents to not registering as RIs, including:

(a) No training for RI registration or examination (Group 8(iv)); and

(b) No experience or inexperience in building maintenance and or refurbishment works (Group 8 (ii)).

The key factors in which a building professional may re-consider registering as RI will be analysed under a separate section.

Table 8.7
Reasons of RI-eligible Building Professionals
for Not Planning to Register as RIs

Group	Description	No. of Response		Adjusted Response	
		No.	Percentage (%)	No.	Percentage (%)
1	Not answered	8	6.7%	0	0.0%
2	Not interested in participating as an RI	36	30.3%	36	32.4%
3	Prefer working in other construction-related professional services than as an RI	16	13.4%	16	14.4%
4	Consider the responsibilities and liabilities of being an RI under the MBIS has not yet clearly defined	16	13.4%	16	14.4%
5	Consider the current remuneration of being an RI not corresponding to the responsibilities and liabilities set forth in the MBIS	8	6.7%	8	7.2%
6	Consider the opportunity cost of providing RI services is higher than other construction-related professional services	4	3.4%	4	3.6%
7	Consider the title of "Inspector" is misconceived as para-professional with low social recognition	8	6.7%	8	7.2%
8	Others				
i	Being civil servant	4	3.4%	4	3.6%
ii	Insufficient experience in building maintenance and/or refurbishment works	3	2.5%	3	2.7%
iii	Consider their professional isn't relevant to provide RI services	3	2.5%	3	2.7%
ii	Lack of training/resources	1	0.8%	1	0.9%
iv	RI's duties is not related to current employment	2	1.7%	2	1.8%
v	Route and qualification for registering as an RI has not yet clearly set	1	0.8%	1	0.9%
vi	Authorised person have already possessed knowledge/ experience way above RI and there is no point to register themselves as RI again	2	1.7%	2	1.8%
vii	The RI examination doesn't suit for selecting the right person to carry out the inspection and suspension works	2	1.7%	2	1.8%
viii	Miscellaneous	5	4.2%	5	4.5%
	Total :	119	100.0%	111	100.0%

8.7 Analysis on Factors Leading to RI Registration for RI-eligible Building Professionals

The key factors which an RI-eligible building professional may re-consider registering as an RI, has been identified in the web-based questionnaire. Out of the total of 119 RI-eligible building professionals, 109 have provided their views (see Table 8.8).

Based on the survey results shown in Table 8.8, 33 respondents (Group 2) (30.8%) advised that they might re-consider registering as RIs if clear setting of responsibilities and liabilities were provided by the Buildings Department. Groups 3 (24.3%) and 4 (8.4%), with a total of 35 respondents (32.7%), might re-consider registering as RIs if the remuneration or opportunity cost of being an RI is improved. It is of paramount importance to note that 15 respondents (14.0%) considered the need to improve social recognition, including re-designation of the name "RI".

Apart from the key factors, two other useful factors given by respondents were as follows:

(a)   Training (Group 6(ii)); and

(b)   Relaxation of maintenance experience (Group 6(iii)).

8.8 Reasons of Current RIs Not Providing Building Inspection and/or Repair Works

As previously advised, 52 out of the total 299 valid respondents in Stage 1 survey were provided by registered RIs. Among them, 30 (57.7%) were practising RIs and 22 (42.3%) were non-practising RIs. Questions were then set to ask the 22 non-practising RIs for their reasons of not providing building inspection and/or repair works services. One respondent (Group 1) did not answer this question. Other responses are detailed in Table 8.9.

The purpose of the above questions was to tackle the reasons for not serving building inspection and/or repair works, despite they were listed in the Inspectors' Registry. Nine Respondents (Group 2, 42.9%) advised that they were currently under full time employment and were not allowed to provide RI services. Five respondents (Group 5, 23.8%) considered that providing other construction-related professional services, rather than RI services, was more satisfying. Two respondents (Group 3, 9.5%) advised that the current remuneration for providing RI services for building inspection and/or maintenance works was not corresponding to the responsibilities set forth by the MBIS.

It is noted with surprise that one respondent (Group 7(i), 4.8%) retired despite his registration as an RI. Lastly, two respondents (Group 7(ii), 9.5%) advised that there were no jobs available to them.

Table 8.8
Analysis of Factors of RI-eligible Building Professionals
for Re-considering Registering as Registered Inspectors (RIs)

Group	Description	No. of Response		Adjusted Response	
		No.	Percentage (%)	No.	Percentage (%)
1	Clear setting of responsibilities and liabilities are to be provided by the Buildings Department of Hong Kong	33	27.7%	33	30.8%
2	When the remuneration of RI is omproved and it corresponds to the responsibilities and liabilities set forth in the MBIS	26	21.8%	26	24.3%
3	When the opportunity cost of providing RI services is lower than other construction-related professional services	9	7.6%	9	8.4%
4	I consider providing other construction-related professional services, rather than RI services, give me more job	15	12.6%	15	14.0%
5	Others				
	i Not interest to re-consider registering as RI	4	3.4%	4	3.7%
	ii Training can be provided	3	2.5%	3	2.8%
	iii Relaxation of intensive maintenance experience by BD	3	2.5%	3	2.8%
	vi After retire	4	3.4%	4	3.7%
	iv Only if it becomes relevant to employment	3	2.5%	3	2.8%
	v Removal of discrimination to non-RI direct related building professions in examination	2	1.7%	2	1.9%
	vi Miscellaneous	5	4.2%	5	4.7%
6	Nil response	12	10.1%	0	0.0%
	Total :	119	100.0%	107	100.0%

All in all, out of the 22 non-practising RIs, 6 might be interested in "re-providing" RI services on building maintenance and refurbishment works. The groups are:

- Group 3 – 2 (9.5%);

- Group 6 – 1 (4.8%)

- Group 7(ii) and (iii) – 3 (14.3%)

- Total = 6 (28.6%)

Table 8.9
Analysis of Reasons of Current Registered Inspector (RIs)
for not Providing Building Inspection and/or Repair works Services

Group	Description	No. of Response No.	No. of Response Percentage (%)	Adjusted Response No.	Adjusted Response Percentage (%)
1	I am currently having full time employment and not allowed to provide part time RI services	9	40.9%	9	42.9%
2	I consider the current remuneration of providing RI services not corresponding to the responsibilities set forth in the MBIS	2	9.1%	2	9.5%
3	I consider the opportunity cost of providing RI services is higher than other construction-related professional services	0	0.0%	0	0.0%
4	I consider providing other construction-related professional services, rather than RI services, give me more job satisfactions	5	22.7%	5	23.8%
5	I consider the working hours of an RI is much longer than other construction-related professional services	1	4.5%	1	4.8%
6	Others				
	i Retired	1	4.5%	1	4.8%
	ii No available jobs	2	9.1%	2	9.5%
	iii MBIS not yet implemented by BD	1	4.5%	1	4.8%
7	Nil response	1	4.5%	0	0.0%
	Total :	22	100.0%	21	100.0%

Adding to the 30 (57.7%) practising RIs in Group 1, the estimated maximum number of possible practising RIs would be 36 (69%).

8.9 Conclusions

From Stage 1 survey, a total of 299 valid responses were received from potential RI-eligible building professionals from the three professional institutions, i.e., the HKIA, HKIE and HKIS. Based on these responses, it is estimated that only 58% of RIs are currently practising in the inspection and maintenance discipline. By applying the same hypothesis, the estimated number of total practising and non-practising RIs would be assessed by using the number of RI registered under the inspector's registry holding by Buildings Department. For example, there were totally 420 RIs as of 22 February 2015 which means only 242 practising RIs were available for implementing the MBIS.

Analysis from Stage 1 survey results revealed that there would be as many as 7,269 potential RI-eligible building professionals in the HKIA, HKIE and HKIS, or alternatively 2,751 RI-eligible building professionals in ARB, ERB and SRB whom were interested or planning to be listed in the Inspectors' Registry of the Buildings Department.

Further questions were made to explore the reason(s) why RI-eligible building professionals did not register themselves as RIs. It was found that around half (46.8%) of the respondents expressed a lack of interest in participating or serving in the field of building inspection and/or repair.

Survey results further revealed that out of the total 107 non-RI respondents, around one-third (30.8%) might re-consider registering themselves as RIs if a clear setting of responsibilities and liabilities is provided by the local authority such as Buildings Department. Another one-third (32.7%) might re-consider registering as RIs if the remuneration or opportunity cost of being RIs is improved. Another 14% of the respondents considered the need of improving social recognition, including re-designation of the title "RI", as the name "inspector" was of para-professional grade in the construction industry. Some respondents suggested that training for attending required professional interview should be provided. Some even suggested that the maintenance experience requirements for the purpose of registering as RIs should be relaxed.

Questions were also raised to the 22 non-practising RI respondents to explore their reason(s) for not providing RI services, despite being listed as RIs in the Inspectors' Registry. Survey results reveal that 43% of the respondents were in fact under full time employment in fields other than building inspection and/or repair services. Another 24% of the respondents preferred to stay in their current practice, largely on new building construction works. The rest, around 29% of the respondents, advised that they might be interested in providing RI services on building inspections and/or repair services when the level of remuneration and/or the marketability of RI services is greatly improved.

References

Buildings Department (2011). *Registration of MBIS Registered Inspectors Begins (APP-7)*. Buildings Department of HKSAR. Available at http://www.bd.gov.hk/english/documents/news/20111229Bae.htm (Accessed on 21 March 2012).

Buildings Department (2012a). *Mandatory Building Inspection Scheme (Pamphlet on MBIS)*. Hong Kong: Buildings Department of HKSAR.

Buildings Department (2012b). *Mandatory Window Inspection Scheme (Pamphlet on MWIS)*. Hong Kong: Buildings Department of HKSAR.

Buildings Department (2013c). *Mandatory Building Inspection Scheme and Mandatory Window Inspection Scheme*. Buildings Department of HKSAR. Available at http://www.gov.hk/en/theme/bf/pdf/Annex1RC.pdf (Accessed on 21 April 2014)

Development Bureau (2011c). *Concern on Maintenance of Aging Buildings arising from the Recent Canopy Collapse Incident in Tuen Mun, Legislative Council Panel on Development* (CB(1)2487/10-11(04), 20 June 2011). Hong Kong: Development Bureau of HKSAR.

Development Bureau (2011d). *Legislative Council Panel on Development—Subsidiary Legislation for Implementation of Mandatory Building Inspection Scheme and Mandatory Window Inspection Scheme* (CB(1)137/11-12(05), 25 October 2011). Hong Kong: Development Bureau of HKSAR.

Hong Kong Institute of Architects (HKIA) (2013). *Looking for Architects*. Hong Kong Institute of Architects. Available at http://www.hkia.net/en/LookingForArchitects/LookingForArchitects_01.htm (Accessed on 1 March 2013)

Hong Kong Institution of Engineers (HKIE) (2013). *Number of Members Advised by Qualification and Membership Board* (e-mail message to Author, 12 March 2013). Hong Kong: HKIE.

Hong Kong Institute of Surveyors (HKIS) (2010). *Letter to Legislative Council— Comments on Building (Amendment) Bill 2010* (16 July 2010). Hong Kong: HKIS.

Hong Kong Institute of Surveyors (HKIS) (2013). *Membership Statistics as at 1 January 2013*. HKIS. Available at http://www.hkis.org.hk/ufiles/memstat-20130101.pdf (Accessed on 1 March 2013)

Legislative Council (2011a). *Paper for the Bills on Buildings (Amendment) Bill 2010, Proposed amendments to include the new building safety initiatives* (LC Paper No. LS62/10-11, 17 May 2011). Hong Kong: Legislative Council of HKSAR.

Legislative Council (2011b). *Panel on Development—Updated Background Brief on Building Safety* (LC Paper No. CB(1)2930/10-11(04), 16 August 2011). Hong Kong: Legislative Council of HKSAR.

Legislative Council (2011c). *Paper on Development—Subcommittee on Building Safety and Related Issues, Meeting on 26 August 2011, Updated Background Brief on Building Safety* (LC Paper No. CB(1)2930/10-11(04), 26 August 2011). Hong Kong: Legislative Council of HKSAR.

Legislative Council (2011d). *Legislative Council Brief—Subsidiary Legislation for Implementation of Mandatory Building Inspection and Mandatory Window Inspection Scheme* (DEVB (PL-CR) 2/15-08), October 2011). Hong Kong: Legislative Council of HKSAR.

Legislative Council (2011e). *Panel on Development—Updated Background Brief on Mandatory Building Inspection Scheme and Mandatory Window Inspection Scheme* (LC Paper No. CB(1)137/11-12(06), 24 October 2011). Hong Kong: Legislative Council of HKSAR.

Legislative Council (2012a). Mandatory building and window schemes. *Official Record of Proceedings* (pp. 15016–15023, 13 June 2010). Hong Kong: Legislative Council of HKSAR.

Legislative Council (2012b). *Subcommittee on Buildings (Amendment) Ordinance 2011 (Commencement) Notice 2012, Building (Inspection and Repair) Regulation (Commencement) Notice and Building (Minor Works)(Amendment) Regulation 2011 (Commencement) Notice* (LC Paper No. CB(1)2026/11-12(01), 29 May 2012). The Government of Hong Kong SAR.

Sing, C. P., Love, P. E. D., and Tam, C. M. (2012). Multiplier model for forecasting manpower demand. *ASCE Journal of Construction, Engineering and Management, 138*(10), 1161–1168.

Sing, C. P., Love, P. E. D., and Tam, C. M. (2014). Forecasting the demand and supply of technicians in construction. *ASCE Management in Engineering, 30*(3), 10.1061/(ASCE) ME.1943-5479.0000227.

CHAPTER 9
Conclusions and Recommendations

9.1 CONCLUSION

The primary objective of the study is to evaluate the supply and demand for RIs for carrying out building inspection and repair works under the Mandatory Building Inspection Scheme (MBIS) which has been implemented on 30 June 2012. A database of private buildings in Hong Kong was consolidated to evaluate the number and size (in terms of number of building units) of private buildings that fell under the MBIS.

An extensive literature review on the population and growth rates, problems of building dilapidations, number of private buildings falling under the regulation of the MBIS, number of potential Registered Inspector (RI)-eligible building professionals in the construction field, and the application and registration rates of RIs was conducted to evaluate the demand for RIs required by the MBIS and the supply of RIs available in the market. A two-stage self-administered questionnaire survey was used. Stage 1, a self-selected web-based survey, analyses the potential number of building professionals who would register as RIs under the MBIS. Stage 2 studies the number of RIs required by efficient implementation of the MBIS.

In light of the rapidly worsening building dilapidation problems and the large number of fatal and serious injury accidents caused, the enactment of the Government to implement the MBIS has obtained fully support from the public. Our primary focus is to evaluate the workforce supply and demand for RIs under the MBIS. Due to the limited information and implementation schedule, Mandatory Window Inspection Scheme (MWIS) is not included in this study.

In order to project the number of aged private buildings falling under the MBIS in the coming 10 years, the rate of ageing of private buildings in the 18 districts was examined. A two-stage web-based questionnaire was administered to building professionals to collect first-hand data in relation to the MBIS. The survey reveals that a large number of building professionals, including architects and engineers, were interested in registering themselves as RIs. Respondents who were already registered as RIs provided the RI-time for the inspection and rectification works under the MBIS. Using the mathematical modeling developed in Chapter 8, it is deduced that about 90 full time RIs are required to conduct the inspection of 2,000 aged buildings per year, which is the MBIS target. If inspection and repair works were implemented at the same time, another 351 RIs will be required. The total number of RIs required should be around 441 when the MBIS is implemented at full speed and in all aspects. It further offers a good reference to the proportion of the estimated number of practising or non-practising RIs currently listed in the Inspectors' Registry of the Buildings Department. By comparing the workforce demand for RIs in 2015 with that in 2025, it is estimated that the total number of full time RIs required to inspect and repair a total of 2,000 buildings under the MBIS will increase from 441 to 470. The rise represents an increase of 6.6% of RI-time and is due to the increase in size (in terms of building unit) of private buildings in the next 10 years.

9.2 RECOMMENDATIONS

9.2.1 Propagation and Education

"Prevention is better than cure" is a motto that deserves our full support. The Development Bureau believes that "if owners can regularly inspect their buildings, identify problems at an early stage and carry out remedial works, accidents can be avoided" (Development Bureau, 2010a). The legislated MBIS should be regarded as the last resort to urge building owners and owners' corporations to maintain their properties in a timely fashion. Along with offering various subsidiaries with quasi-government organisations such as the Hong Kong Housing Society (HKHS) and the Urban Renewal Authority (URA), the Government should consider strengthening education and promoting proper mindsets in the community towards timely maintenance of buildings and properties, in order to encourage owners to take initiative. It will be more useful to construct the mindset and "nurture a building care culture" that timely maintenance of buildings and properties is an obligation of citizens. By so doing, not only may most, if not all, accidents be avoided, the values of their buildings and properties can also be up kept. Education would be vital in nurturing a culture of building in Hong Kong.

9.2.2 Supply of Registered Inspectors (RIs)

Riding on the survey responses, the following are suggested to enhance the supply of RIs:

a) Training

It is noted that the RI registration rates in all 7 disciplines or divisions of the ARB, ERB and SRB, other than the professions of building surveying, are fairly low. According to the survey, a considerable number of respondents who were RI-eligible building professionals commented on the need to receive training for their preparation of professional interviews conducted by the IRC.

b) Relaxation of Maintenance Experience Requirement

While the quality or standard of RIs can be safeguarded by professional interviews, a considerable number of respondents suggested relaxing at least the time length of building inspection, maintenance and/or repair experience set forth in PNAP APP-7.

c) Re-titling the Name of "Inspector"

The survey also reveals that a considerable number of respondents found the name "Registered Inspector" misconceiving and "para-professional", indicating potential low social recognition. In this connection, the Government, policy makers and relevant professional institutions may consider re-naming the profession, especially for the word "inspector", to promote social recognition for this newly formed profession. Some suggestions are proposed herewith—(i) Professional Building Certifier, (ii) Registered Building Certifier or (iii) Authorised Building Certifier.

d) Expanding the Pool of RI-eligible Building Professions

Again, whilst the quality or standard of the can be safeguarded by the professional interviews, the Government, policy makers, and relevant authorities may consider to further expand the pool of other potential eligible building professionals which may include, but not limited to, the building professionals recognised under the Voluntary Building Assessment Scheme (VBAS) and other building professionals like the Royal Institution of Chartered Surveyors (RICS) and the Chartered Institute of Buildings (CIOB) etc.

9.2.3   The Demand of RIs

There is a number of possible ways to ease the shortage of RIs. Along with the proposed solutions to enhance the supply of RIs, the following shall be considered by the Government or policy makers on the demand for the RIs:

a) Integration of Building Classifications and Score Rating System in the Voluntary Building Assessment Scheme (VBAS)

Along with the MBIS, the VBAS operated by the HKHS would gain from the Building Authority that buildings certified by the VBAS will be considered to have fulfilled the MBIS requirements (Hong Kong Housing Society, 2013) and thus exempted from the MBIS for the next inspection cycle (i.e., 10 years). Instead of having an across-the-board 10 years re-inspection cycle, the Government or policy maker may consider, in line with the VBAS, integrating a building classification and score rating system in the VBAS in which the standard re-inspection cycle of 10 years be extended for those buildings achieving high score. The philosophy behind this is that buildings achieving a high score in VBAS are likely to pose less risk to the public when compared with other buildings. Therefore, there is no reason for buildings scoring a high rate to be subject to the same 10 years re-inspection cycle in the MBIS. As a result, this will not only encourage building owners and owners' corporations to participate in the VBAS voluntarily, but also properly maintain their own buildings and properties out of their own initiatives instead of being chosen by the Government.

b) Flowing Numbers of Target Buildings

Currently, a total of 2,000 target buildings is to be selected on a yearly basis under the MBIS. Instead of having an across-the-board and rigid policy of having 2,000 target buildings per year, the Government or policy maker may set a yearly maximum number of target buildings for the MBIS. However, the actual number of buildings under the MBIS may be adjusted by the Government according to the resources available, including, but not limited to, the number of building professionals, RIs, labour, and the economic situation in Hong Kong. The rationale behind is to ensure the quality of inspection and repair works despite changes in the availability of resources, while allowing flexibility in workload within the construction industry. Such regulation will be useful particularly in times when the construction business in Hong Kong is in extreme inflation or depression.

c) Re-distribution of the Scope of Works between RI and RI's Representative

Under Section 6.4 of the Code of Practice (CoP) for the MBIS and MWIS, an RI may appoint and designate a person as his/her representative to provide supervision in a number of activities and the repair of building elements during stages of repair works on his/her behalf (Buildings Department, 2013a). Balancing the risk and the need for professional skills, the Government may review and re-define the scope of works from an RI to his/her representative such that the demands of RI and thence the number of RIs required will be reduced. To balance the risk and the need for professional judgement, in particular items conducted by specialist(s), it is highly suggested only those items in which the RI serves as a witness should be shifted to his/her representative, e.g., conducting pull-off tests, non-destructive testing of welds, CCTV, etc.

d) Reduction of the Scope of Works under MBIS

Likewise, upon balancing the risk on public safety, the Government, policy makers and relevant authorities may also review, re-define and reduce the work item(s) currently included in the MBIS to reduce the inherent risk to the public. Again, this may reduce the demand for the RI's time and thence the number of RIs required.

The recommendations above are by no means exhaustive, but may provide the basis on which a further contribution can be made to measure and improve the performance of the MBIS. The findings presented in this research provide valuable insights which may enable the policy makers, relevant authorities, registration boards and professional institutions to review, formulate and adjust their policies to address the inherent problems of resource constraints associated with the newly implemented MBIS.

References

Development Bureau (2010a). *Government to legislate for mandatory building and window inspection schemes (with videos)* (21 January 2010) [Press Releases]. Hong Kong: Development Bureau of HKSAR.

Buildings Department (2013a). *Target Buildings under Mandatory Building Inspection Scheme (MBIS) and Mandatory Window Inspection Scheme (MWIS) with statutory MBIS/ MWIS Notices served by the Buildings Department.* Buildings Department of HKSAR. Available at https://mwer.bd.gov.hk/MBIS/MBISSearch.do?method=SearchMBIS (Accessed on 6 March 2013)

Hong Kong Housing Society (2013). *Voluntary Building Assessment Scheme—Certificated Buildings.* Hong Kong Housing Society. Available at http://vbas.hkhs.com/en/certified_ buildings/statistics_on_applications_for_vbas_building_certification.php (Accessed on 1 February 2013).

Appendix A
Two-stage Web-based Survey on Building Professional Members

(A1)　The Two-Stage Questionnaire Survey on MBIS and Required RI-time

For reasons of privacy and inaccessibility of contact information of potential respondents, a web-based two-stage questionnaire survey was chosen (Jansen et al., 2007). Stage 1 is a self-selected web-based survey fully supported by the HKIA, HKIE and HKIS; questionnaires were sent by these professional institutions via e-Broadcasts/e-Newsletters to their corporate members of relevant disciplines and divisions. The survey was conducted from 29 August 2012 to 30 October 2012. Respondents, self-selected and happened across the survey in the course of their browsing, were invited to respond to the web-based survey via an external network server named SurveyMonkey®.

With the consent and contact information provided by the respondents in Stage 1, they were then invited to the second round of web-based survey via SurveyMonkey®. It was conducted online between 1 October 2012 and 5 November 2012 to collect workforce data related to RI-time required for the inspection and rectification works under MBIS.

(A2)　Stage 1 Questionnaire Survey on MBIS

The objectives of Stage 1 of the questionnaire survey are to examine the number of RI-eligible building professionals who already registered as RIs and to analyse the potential number of building professionals who would register as RIs under MBIS. Depending on the background and professional qualifications, respondents of the Stage 1 of the questionnaire survey were required to answer questions according to the pre-set logic flow as shown in Figure A.1.

Apart from demographical questions, Stage 1 respondents were required to answer questions on RI registration and practising RI and building maintenance services. For non-RI respondents, questions on circumstances under which they would re-consider registering as RIs were asked. For RI respondents, questions on whether they were currently practising RI or building maintenance services were asked. The objectives of the questions were:

1. To identify the key factors which affect respondents' decision on RI registration;

2. Under what scenario, non-RI respondents would re-consider registering as RIs; and

Figure A.1 Stage 1 Questionnaire Survey Pre-set Logic Flow Diagram

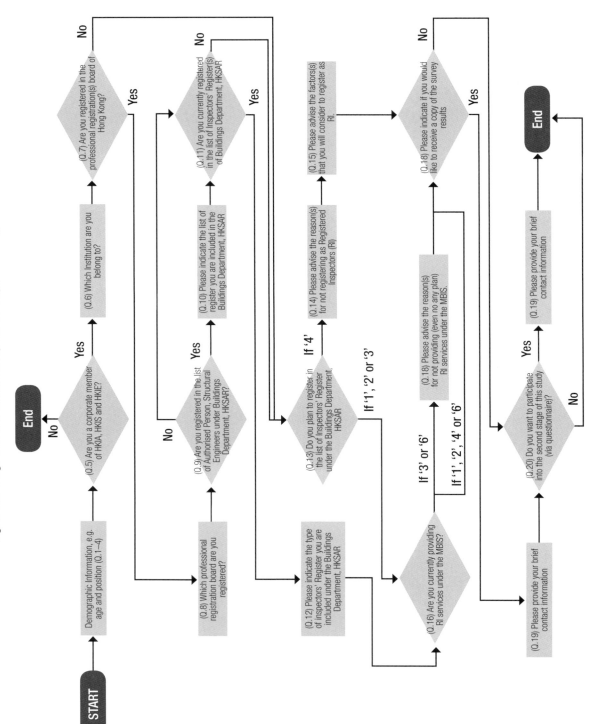

3. To identify the key circumstances under which respondents would provide RI and building maintenance services.

(A2) Stage 2 Questionnaire Survey on Estimated RI Workforce Demand

Stage 2 of the survey was also conducted online via Surveymonkey®. Respondents from Stage 1, who expressed their interest in participating in further studies on MBIS, were invited to provide workforce data related to the inspection and rectification works under MBIS.

To determine the demand for RI workforce under MBIS, Stage 2 respondents were asked to provide the estimated time for inspection and repair works for private buildings in different size groups (in terms of number of building units) as required by MBIS. This demarcation dovetails the prior study on fee schedule of professional services for building inspection under the MBIS conducted by the Urban Renewal Authority (URA, 2012). The survey was designed to divide private buildings, aged 30 years or above into four groups (B_i) according to the number of units in each building. Categories are as follows:

a. Buildings with total no. of units ≤ 50, B_1;

b. Buildings with total no. of units = 51 to 200, B_2;

c. Buildings with total no. of units = 201 to 400, B_3; and

d. Buildings with total no. of units = 401 to 800, B_4.

To estimate the RI-time required at different stages of MBIS, the survey further divided the RI services into "Inspection" and "Rectification and Repair". The "Inspection" stage would deal with the estimated time required by RIs to inspect a building. The "Rectification and Repair" stage would consider the estimated time required by RIs to carry out necessary rectification and repair works according to the Code of Practice (CoP) for MBIS and MWIS published by the Buildings Department (2012f). RI services are further broken down as follows:

Stage 1 — Inspection

RI works covered in this part include:

1. Pre-inspection

2. On-site survey

3. Building diagnosis

4. Inspection report to the BA

Stage 2 — Rectification and Repair

The rectification and repair works under this part are further divided into:

    i.   Pre-Construction Stage

    ii. Construction Stage

    iii. Completion and Certification Stage

i.   Pre-Construction Stage

RI works of this part include:

- Detailed Proposal and Cost Estimation

- Preparation of Tender Document

- Tendering Stage

- Contract Completion

ii.   Construction Stage

RI works of this part include the physical "construction works" complying with the MBIS requirements.

iii. Completion and Certification Stage

RI works of this part included the preparation and submission of the "Completion Report and Certification to the BA" in compliance with the MBIS requirements.

Bibliography

Adams, D., and Hastings, E. M. (2001). Urban renewal in Hong Kong:Transition from development corporation to renewal authority. *Land Use Policy, 18*(3), 245–258.

Booty, F. (2009). *Facilities Management Handbook*. UK: Elsevier.

Buildings Department (2003). Buildings department—Opens public consultation on timely building maintenance (2 November 2003) [Press release]. Hong Kong: Buildings Department of HKSAR.

Buildings Department (2005). Speak up for building safety (27 November 2005) [Press release]. Hong Kong: Buildings Department of HKSAR.

Buildings Department (2006). Submission of views on mandatory building inspection scheme urged (11 January 2006) [Press release]. Hong Kong: Buildings Department of HKSAR.

Buildings Department (2007). Timely maintenance ensures safety (7 January 2007) [Press release]. Hong Kong: Buildings Department of HKSAR.

Buildings Department (2007). Building safety contributes to quality city life (14 October 2007) [Press release]. Hong Kong: Buildings Department of HKSAR.

Caccavelli, D., and Genre, J. L. (2000). Diagnosis of the degradation state of building and cost evaluation of induced refurbishment works. *Energy and Building, 31*(2), 159–165.

Chan, C. M. (2011). Understanding the policy of mandatory building inspection in Hong Kong (Unpublished M.P.A. Dissertation). Hong Kong: City University of Hong Kong.

Chan, H. C. (2013). A study of registered inspectors and aged private buildings in Hong Kong under the mandatory building inspection scheme (Unpublished Engineering Doctorate Dissertation). Hong Kong: City University of Hong Kong.

Consumer Council (2006). *Mandatory building inspection scheme document for consultation.* Consumer Council of HKSAR. Available at http://www.consumer.org.hk/website/ws_en/competition_issues/policy_position/2006031502.html

Development Bureau (2008). Meeting of the Legislative Council Panel on Development —*Progress of Works of the Urban Renewal Authority* (CB(1)1951/07-08(04), June 2008). Hong Kong: Development Bureau of HKSAR.

Development Bureau (2008). Meeting of the Legislative Council Panel on Development —*Review of the Urban Renewal Strategy* (CB(1)1951/07-08(03), June 2008). Hong Kong: Development Bureau of HKSAR.

Development Bureau (2011). Subsidiary legislation to implement mandatory building and window inspection schemes (26 October 2011) [Press release]. Hong Kong: Development Bureau of HKSAR.

Development Bureau (2011). Legislative Council Brief—*Buildings Legislation (Amendment) bill 2011 (DEVB(PL-B) 30/30/122*, 23 November 2011). Hong Kong: Development Bureau of HKSAR.

Development Bureau (2011). *Administration's response to follow-up issues of the meeting held on 17 November 2011, Subcommittee on building (inspection and repair) regulation, Building (Administration)(Amendment) Regulation 2011, Building (minor works)(Amendment) regulation 2011, and Buildings amendment) ordinance 2011 (Commencement) notice 2011* (CB(1)480/11-12(01), November 2011). Hong Kong: Development Bureau of HKSAR.

Development Bureau (2012). LCQ18: Mandatory building inspection scheme (13 June 2012) [Press release]. Hong Kong: Development Bureau of HKSAR. Available at http://www. info.gov.hk/gia/general/201206/13/P201206130430.htm

Government Information Centre (2003). *Public consulted on building management and maintenance* (29 December 2003) [Press release]. Hong Kong: HKSAR. Available at http://www.info.gov.hk/gia/general/200312/29/1229198.htm

Government Information Centre (2011). *Subsidiary legislation to implement mandatory building and window inspection schemes* (22 February 2006) [Press release]. Hong Kong: HKSAR. Available at http://www.info.gov.hk/gia/general/200602/22/ P200602220213.htm

Gregory, T. (2008). Timely inspection a must. *Journal of Building Appraisal, 4*(1), 37–40.

Ha, S. K. (2007). Housing regeneration and building sustainable low-income communities in Korea. *Habitat International, 31*(1), 116–129.

Ho, C. W., , Chau, K. W., Cheung, K. C., Yau, Y., Wong, S. K., Leung, H. F., Lau, S. Y., and Wong, W. S. (2008). A survey of the health and safety conditions of apartment buildings in Hong Kong. *Building and Environment, 43*(5), 764–775.

Ho, K. Y. (2006). Management and maintenance of aged private buildings: Changes in strategies and policies of the HKSAR government (Unpublished MSc Thesis). Hong Kong: University of Hong Kong.

Hong Kong Bar Association (2005). *Public Consultation on Mandatory Building Inspection— Views on the Hong Kong Bar Association*. Hong Kong: Hong Kong Bar Association.

Hong Kong Housing Society (2005). *Building Management and Maintenance Scheme*. Hong Kong Housing Society. Available at http://bmms.hkhs.com/eng/pm_bmm.html

Hong Kong Housing Society (2010). *Housing society renders full support to the mandatory building inspection scheme* (21 January 2010) [Press statement].Hong Kong Housing Society. Available at http://www.hkhs.com/eng/wnew/pdf/pr_20100121.pdf

Hong Kong Housing Society and Urban Renewal Authority (2011). *Integrated Building Maintenance Assistance Scheme (IBMAS)*. Hong Kong Housing Society and Urban Renewal Authority. Available at http://www.hkhs.com/eng/business/pm_ibmas.asp

Hong Kong Housing Society and Urban Renewal Authority (2011). *Integrated Building Maintenance Assistance Scheme (IBMAS)*. Hong Kong Housing Society and Urban Renewal Authority. Available at http://www.hkhs.com/eng/business/pdf/ibmas_leaflet. pdf

Hong Kong Institute of Housing (2012). *The government should launch mandatory building and window inspection schemes as soon as possible and step up promotion and support* (26 April 2012) [Press release]. Available at en.prnasia.com/story/60678-0.shtml, Hong Kong SAR.

Hong Kong Institute of Surveyors (2007). *Revised Draft Guidelines for Building Inspection, Assessment and Rectification Works under the Mandatory Building Inspection Scheme* (Building Surveying Division Chairman's Message). Hong Kong Institute of Surveyors. Available at http://www.hkis.org.hk/hkis/html_bsd/newsroom_chairman_detail.jsp?id=54

Hong Kong Institute of Surveyors (2009). LegCo panel paper on inspectors for MBIS & MWIS (Building Surveying Division Chairman's Message). *Surveyor Times* (February 2009). Hong Kong: Hong Kong Institute of Surveyors.

Hong Kong Institute of Surveyors (2011). Urban renewal in Hong Kong: A community aspiration study (Building Surveying Division Chairman's Message). *Surveyor Times* (April 2011). Hong Kong: Hong Kong Institute of Surveyors.

Hong Kong Institute of Surveyors (2011). Up-dates on building (Amendment) Bill 2010 for MBIS (Building Surveying Division Chairman's Message). *Surveyor Times* (April 2011). Hong Kong: Hong Kong Institute of Surveyors.

Hong Kong Institute of Surveyors (2011). Problems in aged building: Subdivided flats (Building Surveying Division Chairman's Message). *Surveyor Times* (June 2011).Hong Kong: Hong Kong Institute of Surveyors.

Hong Kong Housing Department (2013). *Independent Checking Unit—Building Control, Corporate Plan 2008/09.* Housing Department of HKSAR. Available at http://www.housingauthority.gov.hk/hdw/en/aboutus/publications/corporateplan0809/page4.html

Housing, Planning and Lands Bureau (2002). *Review of the Pilot Coordinated Maintenance of Building Scheme* (LC Paper No. CB(1)2488/01-02, 6 September 2002). Hong Kong: Housing, Planning and Lands Bureau of HKSAR.

Housing, Planning and Lands Bureau (2003). *Public Consultation on Building Management and Maintenance* (CB(1)674/03-04, 29 December 2003). Hong Kong: Housing, Planning and Lands Bureau of HKSAR.

Housing, Planning and Lands Bureau (2003). *Public Consultation Paper on Building Management and Maintenance* (29 December 2003). Hong Kong: Housing, Planning and Lands Bureau of HKSAR.

Housing, Planning and Lands Bureau (2005). *Report on the Public Consultation on Building Management and Maintenance* (CB(1)768/04-05(01), January 2005). Hong Kong: Housing, Planning and Lands Bureau of HKSAR.

Housing, Planning and Lands Bureau (2005). *Building Management and Maintenance— Public Consultation on Mandatory Building Inspection* (Leaflet). Hong Kong: Housing, Planning and Lands Bureau of HKSAR.

Housing, Planning and Lands Bureau (2006). *Building Management and Maintenance— Public Consultation on Mandatory Building Inspection—Full Consultation Paper.* Hong Kong: Housing, Planning and Lands Bureau of HKSAR.

Housing, Planning and Lands Bureau (2006). *Report on the Public Consultation on Mandatory Building Inspection—Highlights*. Hong Kong: Housing, Planning and Lands Bureau of HKSAR.

Housing, Planning and Lands Bureau (2007). *Public Consultation on Mandatory Building Inspection—Legislative Council Panel on Planning, Lands and Works* (CB(1)1643/06-07(03), 22 May 2007). Hong Kong: Housing, Planning and Lands Bureau of HKSAR.

Lai, W. C., Chau, K. W., Ho, C. W., and Lorne, T. F. (2005). A "Hong Kong" model of sustainable development. *Property Management, 24*(3), 251–271.

Lam, C. M. (2007). Evaluation of the role of property manager in implementation of the mandatory building inspection scheme (Unpublished MSc Thesis). Hong Kong: University of Hong Kong.

Langston, C., Wong, K. W., Hui, C. M., and Shen, L. Y. (2008). Strategic assessment of building adaptive reuse opportunities in Hong Kong. *Building and Environment, 43*(10), 1709–1718.

Mak, K. W. (2008). Management and maintenance of building: Strategy to solve long standing building problem in Hong Kong (Unpublished MSc Thesis). Hong Kong: University of Hong Kong.

The Government of Hong Kong SAR (2007). *Views to be sought on building maintenance* (2 November 2003) [Press release]. Hong Kong: Infrastructure & Logistics, HKSAR. Available at http://sc.info.gov.hk/gb/www3.news.gov.hk/isd/ebulletin/en/category/infrastructureandlogistics/031102/html/031102en06002.htm

The Government of Hong Kong SAR (2006). *Building inspection details ready next year—infrastructure & logistics* (24 November 2006) [Press release]. Hong Kong: HKSAR. Available at http://www.news.gov.hk/en/category/infrastructureandlogistics/061124/html/061124en06004.htm

The Government of Hong Kong SAR (2007). *Mandatory building inspection on the way—infrastructure & logistic* (22 May 2007) [Press release]. Hong Kong: HKSAR. Available at http://www.news.gov.hk/en/category/infrastructureandlogistics/070522/html/070522en06004.htm

Sing, C. P. (2012). Development of sustainable manpower planning system for the construction industry (Unpublished PhD thesis). Australia: School of Built Environment, Curtin University.

Tang, C. M., Wong, C. W. Y., Leung, A. Y. T., and Lam, K. C. (2006). Selection of funding schemes by a borrowing decision model: A Hong Kong case study. *Construction Management and Economics, 24*(4), 349–365.

Tam, C. M., Tong, K. L., and Sing, C. P. (2011). *Manpower Research in Hong Kong*. Unpublished Consultancy Report. Hong Kong: Hong Kong Construction Industrial Council, HKSAR.

Tam, C. M., Tong, K. L., and Sing, C. P. (2011). *Mobility of Construction Workers*. Unpublished Consultancy Report. Hong Kong: Hong Kong Construction Industrial Council, HKSAR.

The Royal Institution of Chartered Surveyors (2012). *RICS HK urges the government to speed-up implementation of mandatory building safety inspection scheme* (24 August 2009) [Press release]. Hong Kong: The Royal Institution of Chartered Surveyors, HKSAR. Available at http://www.prnasia.com/story/22733-0.shtml.

The Royal Institution of Chartered Surveyors (2012). *RICS' five recommendations to government to speed up mandatory inspection plan and minimize damage to human lives* (2 February 2012) [Press release]. The Royal Institution of Chartered Surveyors. Available at http://www.ricsasia.org/newsDetail.php?id=157&regionID=0

The Royal Institution of Chartered Surveyors (2012). *RICS introduces best practice on dilapidations* (25 February 2010) [Press release]. The Royal Institution of Chartered Surveyors. Available at http://www.ricsasia.org/newsDetail.php?id=161&regionID=0